Military Innovation in Small States

This book provides a comprehensive assessment of the global diffusion of the Revolution in Military Affairs (RMA) and its impact on military innovation trajectories in small states.

Although the RMA concept has enjoyed significant academic attention, the varying paths and patterns of military innovation in divergent strategic settings have been overlooked. This book seeks to rectify this gap by addressing the broad puzzle of how the global diffusion of RMA-oriented military innovation – the process of international transmission, communication, and interaction of RMA-related military concepts, organizations, and technologies – has shaped the paths, patterns, and scope of military innovation of select small states. In a reverse mode, how have selected small states influenced the conceptualization and transmission of the RMA theory, processes, and debate? Using Israel, Singapore, and South Korea as case studies, this book argues that RMA-oriented military innovation paths in small states indicate predominantly evolutionary trajectory, albeit with varying patterns resulting from the confluence of three sets of variables: (1) the level of strategic, organizational, and operational adaptability in responding to shifts in the geostrategic and regional security environment; (2) the ability to identify, anticipate, exploit, and sustain niche military innovation – select conceptual, organizational, and technological innovation intended to enhance the military's ability to prepare for, fight, and win wars; and (3) strategic culture. While the book represents relevant empirical cases for testing the validity of the RMA diffusion hypotheses, from a policy-oriented perspective, this book argues that these case studies offer lessons learned in coping with the security and defence management challenges posed by military innovation in general.

This book will be of much interest to students of military innovation, strategic studies, defence studies, Asian politics, Middle Eastern politics, and security studies in general.

Michael Raska is Assistant Professor at the S. Rajaratnam School of International Studies, Singapore, and has a PhD from the National University of Singapore.

Cass Military Studies

Military Innovation in Small States

States

Creating a reverse asymmetry

Michael Raska

Routledge
Taylor & Francis Group

LONDON AND NEW YORK

First published 2016 by Routledge

2 Park Square, Milton Park, Abingdon, Oxfordshire OX14 4RN
52 Vanderbilt Avenue, New York, NY 10017

Routledge is an imprint of the Taylor & Francis Group, an informa business

First issued in paperback 2020

British Library Cataloguing-in-Publication Data
A catalogue record for this book is available from the British Library

Library of Congress Cataloging-in-Publication Data
Names: Raska, Michael.
Title: Military innovation in small states: creating a reverse asymmetry /
Michael Raska. Description: Abingdon, Oxon; New York, NY: Routledge is an
imprint of the Taylor & Francis Group, an Informa business, [2016] |
Series: Cass military studies | Includes bibliographical references and index.
Identifiers: LCCN 2015021499 | ISBN 9781138787230 (hardback)
Subjects: LCSH: Military art and science—Case studies. | Military policy—Case
studies. | Strategic culture—Case studies. | Singapore. Armed Forces. | Israel.
Tseva haganah le-Yiâsra®el. | Korea—Armed Forces.
Classification: LCC U102 .R25 2016 | DDC 355/.0335—dc23
LC record available at http://lccn.loc.gov/2015021499

ISBN: 978-1-138-78723-0 (hbk)
ISBN: 978-0-367-66861-7 (pbk)

Typeset in Times
by Book Now Ltd, London

Contents

Figures

Acknowledgments

This book has evolved throughout my studies, research, and experiences in South Korea, Israel, and Singapore. In all three states, strategic and defense studies in the context of public policy have been of crucial importance, given the many critical factors that have historically shaped their military and security situation. Accordingly, understanding the complexities in the continuity and change of their security paradigms, defense strategies, operational concepts, and evolving security challenges cannot be accomplished without the close consultation of select policy practitioners, military officers and soldiers, defense analysts, leading academics, and journalists who have devoted their professional lives to ensure the continued existence and security of their nation-states. This book thus inevitably treads along signposts and paths mapped out by others who have provided me with insightful comments, thoughts, and observations that sharpened my understanding of the security conceptions in the select states. While any errors in this book are mine, and mine alone, the following lines are dedicated to a number of special individuals for their undivided support, encouragement, expertise, and trusted advice.

While many remain anonymous, special thanks go to Chung Min Lee at the Graduate School of International Studies at Yonsei University in Seoul, who has been at the conceptual genesis of this book, suggesting the idea for a comparative study of Israeli and South Korean air power strategies, as a way to enhance the emerging air power studies in Korea. Richard Bitzinger, Bernard F. W. Loo, Pascal Vennesson, Collin Koh Swee Lean – my mentors, colleagues, and friends at the S. Rajaratnam School of International Studies, Nanyang Technological University – for their guidance and insights on the trajectories of military innovation in Singapore. Darryl Jarvis, Michael Evans, and Gil Merom for their detailed review and comments on the initial draft. In Israel, I owe a great debt of gratitude to Dima Adamsky, Efraim Inbar, Isaac Ben-Israel, Ehud Eiran, Itai Brun, Ilan Mizrahi, Giora Eiland, Deganit Paikowsky, Ram Levi, Arnon Eshel, Amir Horkin, Iris, and Meytal Nasie. Ultimately, my greatest debt – and one impossible to specify – is to my family Christine, Jiri, Christian Raska, and especially to my partner Anita Tng, for inspiring me to believe in my global intellectual endeavors.

About the author

Michael Raska is Assistant Professor in the Military Transformations Program at the S. Rajaratnam School of International Studies (RSIS), Nanyang Technological University in Singapore. His research interests include East Asian security and defense, including theoretical and policy-oriented aspects of military innovation, force modernization trajectories, information conflicts, and cyber-warfare. He is the co-editor of recently published volume titled *Security, Strategy and Military Change in the 21st Century Cross-Regional Perspectives* (Routledge 2015). He has contributed chapters to edited volumes in cooperation with the National Bureau of Asian Research; the Project on the Study of Innovation and Technology in China at the Institute on Global Conflict and Cooperation, UC-San Diego; and the RSIS. His academic writings include journal articles published by the *Korea Journal of Defense Analysis*, *Pointer – Journal of Singapore Armed Forces*, and *Asian Journal of Public Affairs*. His policy-oriented reports and strategic assessments have been published by the *National Interest*, *Project Syndicate*, *Defense News*, *The Diplomat*, *East Asia Forum*, *RUSI Newsbrief*, *RSIS Commentaries*, and select newspapers worldwide.

At the RSIS, Dr Raska is currently teaching a graduate course on Information Conflicts and Computer Network Operations. Previously, he has taught at the Goh Keng Swee Command and Staff College (SAF Campaign and War Studies Course) and the Lee Kuan Yew School of Public Policy (International Security). His research experiences include visiting fellowships at GSIS Yonsei University, Pacific Forum CSIS, and Samsung Economic Research Institute. He is an alumnus of the Columbia/Cornell University Summer Workshop on Analysis of Military Operations and Strategy (SWAMOS 2012); the Philip Merrill Center for Strategic Studies Workshop at Basin Harbor (2013); and the SAIS/Hertog Summer Study Program (2014). He holds a BA in international studies from Missouri Southern State University (2000), an MA in international relations from the Graduate School of International Studies at Yonsei University (2002), and a PhD in public policy (2012) from the Lee Kuan Yew School of Public Policy, where he was a recipient of the NUS President's Graduate Fellowship.

Abbreviations

3G SAF	Third-Generation Singapore Armed Forces
ACMS	Advanced Combat Man System
ALB	AirLand Battle
AMDA	Anglo-Malayan Defense Agreement
APS	Active Protection System
ASEAN	Association of Southeast Asian Nations
ATGM	Anti-Tank Guided Missile
AWACS	Airborne Warning and Control System
BCT	Brigade Combat Team
BMS	Battlefield Management System
BVR	Beyond Visual Range
C4I	Command, Control, Communications, Computers, and Information
C4ISTAR	Command, Control, Communications, Computers, Intelligence, Surveillance, Targeting, and Reconnaissance
CCRP	Command and Control Research Program
CEC	Cooperative Engagement Capabilities
CENTCOM	Central Command
CFC	Combined Forces Command
CMS	Combat Management System
COG	Center of Gravity
DAP	Digital Army Program
DBK	Dominant Battlespace Knowledge
DBM	Dominant Battle Management
DIB	Defense Industrial Base
DMZ	Demilitarized Zone
DOD	Department of Defense
DRP	Defense Reform Plan
DRTO	Defense Research and Technology Office
DSO	Defense Science Organization
DSTA	Defense Science and Technology Agency
EBO	Effects-Based Operations
FAD	Forward Active Defense
FCS	Future Combat System

FEBA	Forward Edge of Battle Area
FOFA	Follow-on Forces Attack
FPDA	Five-Power Defense Arrangement
FSD	Future Systems Directorate
FSTD	Future Systems and Technology Directorate
GCC	Ground Corps Command
GFC	General Forces Command
GS	General Staff
HADR	Humanitarian and Disaster Relief
IAF	Israel Air Force
IAS	Integrated Advanced Soldier
IDF	Israel Defense Forces
IED	Improvised Explosive Device
IKC2	Integrated Knowledge-based Command and Control
ISR	Intelligence, Surveillance, and Reconnaissance
JFCOM	Joint Forces Command
KeDS	Knowledge-enabled Decision Superiority
KPA	Korean People's Army
LIC	Low-Intensity Conflicts
MBT	Main Battle Tank
MCZ	Military Control Zone
MINDEF	Ministry of Defense (Singapore)
MMI	Major Military Innovation
MND	Ministry of National Defense
MNF	Malayan Naval Forces
MPR	Military Participation Ratio
MR	Military Revolution
MTR	Military-Technical Revolution
MWS	Major Weapon Systems
NCW	Network-centric Warfare
NDP	National Defense Panel
NIE	National Intelligence Estimate
NSF	National Servicemen
OFT	Office of Force Transformation
ONA	Office of Net Assessment
OOTW	Operations Other Than War
OPG	Operational Maneuver Groups
OSD	Office of Secretary of Defense
OTRI	Operational Theory Research Institute
OTS	Off-the-shelf
PBA	Pervasive Battlespace Awareness
PGM	Precision Guided Munitions
PSO	Peace Support Operations
QDR	Quadrennial Defense Review

R&D	Research and Development
RDE&T	Research, Development, Evaluation, and Testing
RMA	Revolution in Military Affairs
ROK	Republic of Korea
ROKA	Republic of Korea's Army
RPV	Remotely Piloted Vehicle
RSAF	Republic of Singapore Air Force
RSN	Republic of Singapore Navy
RUK	Reconnaissance-Strike Complexes
SAF	Singapore Armed Forces
SAFTI	Singapore Armed Forces Training Institute
SAM	Surface-to-Air Missile
SBU	Superior Battlespace Understanding
SIR	Singapore Infantry Regiment
SLOC	Sea Lines of Communication
SOD	Systemic Operational Design
SSM	Surface-to-Surface Missile
STR	Stability, Transition, and Reconstruction
TD	Total Defense
TW-PS	Theatre-Wide Precision Strike
UA	Unit of Action
UACV	Unmanned Aerial Combat Vehicle
UAV	Unmanned Aerial Vehicle
USAF	United States Air Force
USFK	United States Forces Korea
USMC	United States Marine Corps
USN	United States Navy
WMD	Weapons of Mass Destruction

1 Introduction

Debating military innovation and small states

One of the quintessential threads in strategic studies has centered on the debate on the sources, paths, and patterns of military change, learning, and innovation. In particular, why and how militaries change, when, how, and why do militaries succeed or fail in adopting and adapting to new ways of war? In 1984, Barry Posen examined why and how military organizations innovate in a comparative study of interwar military doctrines of Great Britain, France, and Germany (Posen 1984). Posen's work triggered new debates in strategic studies, focusing on the causes, catalysts, conditions, and trajectories of military innovation. In particular, when, how, and why do military organizations innovate? What are the key variables, internal or external, that determine the paths and patterns of military innovation? Why do select military organizations succeed or fail in pursuing military innovation, under what conditions? How do major military innovations (MMIs) emerge, diffuse, and shape or change the international security environment relative to evolutionary, sustaining innovations? The nearly three-decade-long debate generated a number of theories with varying explanations of key drivers, patterns, processes, and outcomes of military innovation. Nearly all, however, start from the basic assumption that propensity of change in military organizations – institutionalizing new approaches to warfare – is neither probable nor simple.

Indeed, military innovation studies often point toward three types of barriers to innovation: organizational rigidity, bureaucratic politics, and organizational culture (Davidson 2011, 11). According to organizational theory, for example, military organizations function as large, complex, and functionally specialized bureaucracies, placing a premium on predictability, stability, and certainty (Posen 1984, 46). Coupled with institutionalized standard operating procedures, core missions, goals, strategies, and structural norms, these factors reinforce military stagnation. Similarly, bureaucratic politics theory blames uniformed leaders, seeking to promote the importance of their organizations, maximize resource allocation, and preserve their distinct organizational roles, missions, and capabilities (Halperin 1974). In other words, military organizations tend to protect their self-interests in an environment of scarce resources, which strengthens their risk aversion and resistance to change (Isaacson, Layne, and Arquilla 1999). Therefore, innovations that pose no threat to the organization's mission, resources, autonomy, or essence are readily adopted, while those innovations perceived as a major threat may be strongly resisted (Goldman and Mahnken 2004).

In contrast, culturalist theorists view barriers to innovation embedded in particular strategic as well as organizational cultures. Williamson Murray defines strategic culture as "ethos and professional attributes, both in terms of experience and intellectual study, which contribute to a common core understanding of the nature of war within military organizations" (1999, 27). While strategic culture shapes ideas about the use of force within a state through their historical experiences and lessons learned, organizational culture comprises of identities, norms, and values that have been internalized by military organizations in their regulations, training, routines, and practice. Strategic culture thus conditions the preconceived notions on the character and conduct of warfare and serves both as an enabler or constraint in the effectiveness of organizational learning, adaptation, and innovation (Eisenstadt and Pollack 2001). Ultimately, pursuing and managing military innovation does not arise in a vacuum – the range of cultural, bureaucratic, and organizational barriers to military innovation are amplified by diverse contextual variables that include intrinsic geostrategic, political, economic, and operational variables shaping a state's ability to generate military power (Isaacson, Layne, and Arquilla 1999, 9).

Throughout history, military innovation has thus reflected a seemingly perennial paradox (Farrell and Terriff 2002): on one hand, military organizations have been traditionally resistant to change, preserving tried-and-tested strategies and structures to foster continuity amid the prevailing process of institutionalization and the fact that the cost of error in the face of ubiquitous strategic uncertainty may be exceedingly high (Alberts 1996). On the other hand, militaries have recognized that failure to achieve advances in the ways and means of war may result in defeat and thus are motivated to seek developments that could revolutionize military operations (Pierce 2004). During the era of Napoleonic warfare, Carl von Clausewitz envisioned military innovation as a linear process of emulation and adaptation of effective practices on the battlefield: "If, in warfare, a certain means turns out to be highly effective, it will be used again; it will be copied by others and become fashionable, and so, backed by experience, it passes into general use and is included in the theory" (1984, 171). While Clausewitz did not explain how military learning may transcend from the battlefield into organizational practices, or why select armies learn while others do not, he noted that armies have at least three pathways to learn – historical examples (of self and others), personal battlefield experience, and the experience of other armies (Davidson 2011, 9).

Since then, historical and empirical evidence has shown a much broader variation in the sources, paths, and patterns of military innovation (Goldman and Eliason 2003): from development of new or different instruments (technology), practices (doctrines and operational concepts), to formation of new organizational force structures (Cheung, Mahnken, and Ross 2011, 77–80). Projecting the barriers to military innovation through different theoretical lenses yields different arguments on the appropriate mechanisms to overcome them. The ongoing debate can be structured into four major contending schools of thought: (1) civil–military conflict, (2) intraservice conflict, (3) interservice conflict, and (4) organizational culture (Grissom 2006, 907). The first three frameworks view MMI or disruptive innovation as a result of conflicts in decisive relationships – between civilian and military leaders, within services, or between services, while incremental innovation results from suppressing such conflicts. Barry Posen's *civil–military competition* framework, integrates seemingly contrasting propositions of structural realist theory (balance of power) and organizational theory, arguing that external threats and civilian intervention are primary determinants of military innovation (Posen 1984). Posen views military organizations resisting change by design – as rational, functional, specialized bureaucracies, militaries have institutionalized vested interests to preserve their autonomy, size, and wealth while reducing operational uncertainties. As a result, military innovation can proceed through a direct combat experience with a new technology, major failure on the battlefield, and, most importantly, external civilian intervention forcing change in the military decision-making processes. In this context, civilian intervention is conditioned by the perennial insecurity and threat perceptions shaped by the anarchic structure of the international system. Civilian leaders interpret continuity and change in the security environment and react by modifying internal organizational, bureaucratic, and military commitments. When threat perceptions are low, civilian intervention and military innovation are incremental; when security threats are high, civilian leaders have a greater incentive to intervene and essentially impose major changes on the military. According to Posen (1984), the pathways for such intervention can be either direct or indirect, channeled through a link between civilian leaders and select "maverick" officers, who provide the civilian leadership with necessary military expertise and stimulate innovation within military organizations.

In contrast, proponents of *intra-service competition* model, such as Stephen Peter Rosen, argue that military innovation can be facilitated internally – between branches of the same service, without external civilian interference that often fails (Rosen 1991). According to Rosen, peacetime military innovation emanates from an ideological struggle for power between established and reform-oriented branches within a single service over competing visions of future warfare. Innovation diffuses through internal structural changes, which are gradual and evolutionary, altering the distribution of organizational power among competing organizational factions or subgroups and their preferences. Rosen argues that when intraservice competition between branches is high, innovation accelerates as a result of each branch trying to dominate the other. The essential mechanism for innovative military change is the determination and success of visionary senior officers or "product champions" advocating a new vision of future warfare and searching for allies and resources in reform-minded junior officers tasked to develop and experiment with innovative operational concepts, tactics, and techniques. In order to spearhead innovation, the senior officer "mavericks" open "promotional pathways" such as the establishment of new arms or branches of service for select junior officers, protecting them from political threats and internal struggles within the service, gradually enabling them to foster innovation, and ultimately empowering them to dominate the service. The process may take up to a decade, depending on the rate of junior officers' promotions (Rosen 1991).

The third school of thought centers on *inter-service competition*, which has evolved from studies by Vincent Davis, Bradd Hayes, Owen Cote, and others (Davis 1967; Hayes and Smith 1994; Cote 1996). Its proponents acknowledge that if civilian intervention or intraservice competition may selectively accelerate military innovation by offering solutions for particular strategic or operational problems, so can interservice relations also moderate innovative effects through competitive and cooperative patterns. However, the dynamics of interservice conflict – between the military services within a state for new roles, missions, and resources – has a greater effect on innovation and can act alone and independently even in the presence of strong civilian and intraservice opposition (Cote 1996, 13–86). This is because military organizations are driven by professional ethos to provide security for the state under conditions of great uncertainty, which stimulates debate and amplifies both competition and cooperation between the services for scarce defense resources. Such interservice conflicts – that is, over the development of a new doctrine for the use or integration of a new technology, will be more intense and openly politicized than intraservice conflicts, resulting in greater spill-over effects into political decision-making arenas (Cote 1996, 91–94).

The fourth school of thought rejects military innovation as a result of conflicting relationships; it explains sources of military innovation through differences in *strategic and organizational cultures* – that is, distinctive, consistent, and persistent views on how states and their military organizations think about warfare. According to Dima Adamsky, "different cultures think differently about military innovation and produce various types of doctrinal outcomes from the same technological discontinuity" (2010, 1). In this view, military innovation is conditioned by different national "cognitive styles" – that is, strategic preferences, perceptions, ideas and knowledge, techniques and professional attributes and patterns of habitual behavior acquired over time within members of national strategic community. Theo Farrell defines military culture as "comprising those identities, norms and values that have been internalized by a military organization and frame the way the organization views the world, and its role and functions in it" (2008, 783). In other words, both cultural norms and professional traditions set the context for military innovation, fundamentally shaping organizational choices, preferences, and reactions to technological and strategic opportunities (Farrell 1998). Three select mechanisms may define this process: (1) senior leaders may plan and direct military organizations toward innovation through changes in strategic thought; (2) external shocks can trigger a process of cultural change; and (3) cross-national cultural norms can shape the paths and patterns of military emulation. Similarly, Elizabeth Kier (1997) argues that organizational culture shapes the direction and character of strategic thought of military leaders and, in doing so, state's choices in military doctrine and innovation. At the same time, organizational culture can "play a critical role in determining how effectively organizations can learn from their own experiences" (Nagl 2002, 11). In short, strategic and organizational cultures can serve as both enabler and constraint of military innovation.

In this debate, however, the literature on military innovation has portrayed innovation largely through the argument of *major military change* by great powers in relation to existing ways of war. Stephen Peter Rosen, for example, conceptualized military innovation in the context of major "change that forces one of the primary combat arms of a service to change its concepts of operation and its relation to other combat arms, and to abandon or downgrade traditional missions. Such innovations involve a new way of war, with new ideas of how the components of the organization relate to each other and to the enemy, and new operational procedures conforming to those ideas" (1988, 134). Rosen (1991) differentiated between MMIs and technological innovations, with MMIs further subdivided into *peacetime and wartime processes*. Theo Farrell and Terry Terriff also distinguished MMIs, or "change in the [organizational] goals, actual strategies, and/or structure of a military organization," and minor military innovations, or "changes in operational means and methods (technologies and tactics) that have no implications for organizational strategy or structure" (2002, 5). More recently, Michael Horowitz equated MMIs as "major changes in the conduct of warfare, relevant to leading military organizations, designed to increase the efficiency with which capabilities are converted to power" (2010, 22). Dima Adamsky too focused on disruptive military innovations through the lens of military-technical revolutions (MTRs) or revolutions in military affairs (RMAs), when "new organizational structures together with novel force deployment methods, usually but not always driven by new technologies, change the conduct of warfare" (2010, 1). Inherent in the above definitions has been a *process* of radical/disruptive change, a large-scale, RMA-oriented innovation defined by the synergy of technological change, military systems development, operational innovation, and organizational innovation (Krepinevich 1994, 30–43).

The debate on what constitutes military revolutions (MRs), RMAs, and MTRs in the broader context of military innovations has carved a significant path in strategic studies over the past two decades. Notwithstanding the often differing definitions and schools of thought that have evolved about each conceptual term, all have in essence reflected a "high strategic concept" synonymous with a "discontinuous" or "disruptive" military innovation in the character and conduct of warfare (Gray 2002, 1–27). Since its early inceptions as a MTR theory by Soviet strategic thinkers in the early 1980s to the broader contours of the *U.S. Defense Transformation* in the early 2000s, the underlying visions of future warfare have been anchored in an ongoing paradigm shift from the "industrial-age" toward twenty-first-century "information-age warfare" (Toffler and Toffler 1993, 27–64). In particular, three main arguments have defined the RMA drive: (1) the application of new information technologies into a significant number of military systems coupled with innovative operational concepts and organizational adaptation will fundamentally alter the character and conduct of warfare by producing a dramatic increase in the combat potential and overall military effectiveness (Krepinevich 1994, 30–43). (2) Attaining qualitatively new levels of military effectiveness that transcends marginal improvements will essentially mitigate the widening spectrum of security challenges of the twenty-first century, stipulated by the convergence of conventional, low-intensity, asymmetrical, and non-linear types of conflict (Cohen 1996; Mazarr 1994; Hundley 1999; Murray and Knox 2001). (3) States and military organizations adopting RMA-oriented concepts, advanced defense technologies, and relevant force structures will possess a considerable strategic advantage over those that do not (Mahnken and FitzSimonds 2003; Bitzinger 2008; Loo 2009).

At the forefront of the RMA-oriented conceptual, technological, and organizational military innovations over the past two decades has been the United States, with the world's most sophisticated defense industrial base (DIB), resources, organizational and technological capacity to implement and actively exploit new technologies and concepts as new ways of war. The United States has taken the lead in conceptualizing visions of future wars while developing the next generation of precision-guided munitions (PGMs), intelligence, surveillance, and reconnaissance (ISR) platforms, command, control, communications, computers, and information (C4I) systems, space-based intelligence assets, cyber capabilities, and integrating them with innovative "network-centric" operational concepts. The integration of "sensor to shooter" systems as "force multipliers," including wide area of electronic sensors for long-range, all-weather target detection and acquisition, has been projected to enable near real-time *situational awareness* of the battlefield and, in doing so, mitigate the adverse effects traditionally synonymous with the *fog of war* – or the pervasive nature of uncertainty, ambiguity, fear, and friction of battle (Alberts, Garstka, and Stein 2000). The U.S. military would be then able to conduct rapid, stand-off, and precision strikes beyond the reach of enemy's defenses. In theory, this would significantly mitigate the scope and magnitude of collateral damage, shorten the duration of conflicts, and also minimize combat casualty rates traditionally associated with high-intensity conventional wars. In doing so, U.S. forces would yield a *decisive military advantage* relative to preexisting military capabilities, while shortening the costs and duration of conflicts, and, ultimately, ensure the defeat of any adversary on their terms (FitzSimonds and Van Tol 1994, 24–31).

These varying conceptual, technological, and organizational innovations under the umbrella of RMAs and subsequently defense transformation have been consistently tested, adapted, and debated as a "new way of war" in key conflicts and military campaigns of the U.S. forces – from the Persian Gulf War (1991), through the Air War in Kosovo (1999), and, subsequently, the protracted wars in Iraq and Afghanistan (2003–2010). While the intellectual thrust on what defines and drives the "American" defense transformation, its templates, and key operational terms has changed throughout these conflicts, the underlying strategic aim has remained intact: to transform the U.S. military into a more agile, mobile, integrated, and lethal force, capable of directly "shaping" the strategic environment and defeating any existing or future adversary in line with U.S. global geopolitical aspirations and foreign policy interests. With its ambitious and wide-reaching aims and scope, however, the American vision of defense transformation has ignited intense *policy debates* in the United States concerning its validity, applicability, and utility. Drawing lessons learned from the challenges and operational demands in the counter-insurgency conflicts in Afghanistan and Iraq, the priorities, resources, and focus in the U.S. military establishment during these conflicts shifted from pursuing relatively open-ended and broad defense transformation to tackling insurgent strongholds, sectarian violence, and immediate tactical threats such as improvised explosive devices or IEDs (Adams 2008). The United States and its allies became confronted by a wide spectrum of political, socioeconomic challenges of a non-linear conflict that it neither was prepared for nor anticipated (Shimko 2010, 203). Both conflicts also raised questions about the effectiveness of new advanced weapons and technologies such as UAV drones in winning the "hearts and minds" of the people (Farrell, Terriff, and Osinga 2010). In short, the key questions powering the "American RMA debate" over the past two decades have shifted from (1) the definition, metrics, and impact of the revolution, (2) the pace, direction, and cost of technology in warfare, to (3) its relevance amid changes in the sources and character of security threats. With the increasing sophistication, and more importantly costs, in the development of advanced military technologies and R&D programs, many questions also emerged regarding the affordability, feasibility, and desirability of a comprehensive defense transformation. Over the years, the most pressing questions included: is there an RMA, and if so, what does it mean – what is new or revolutionary? Is it a process signifying a real "disruptive" shift in warfare or a mere continuation of technological progress and modernization of armed forces? If the RMA indeed stipulates a paradigm shift in warfare, what are the defense resource allocation priorities, force structure requirements and procurement needs in the context of future defense planning? Is there a need to pursue the RMA? And ultimately, how effective is the RMA in dealing with the complexity of the security challenges of the twenty-first century, characterized by the increasing volatility and uncertainty through the convergence of traditional security threats and non-linear/asymmetric threats?

The changing debate projected a range of contending theoretical perspectives in the U.S. strategic thought, shaping differing views and beliefs about the use of force in the twenty-first century. Accordingly, with the intellectual thrust on the RMA-oriented military innovation debate focusing on the United States, there has not been a substantial research into its diffusion paths and patterns in divergent strategic settings, particularly its impact on military innovation trajectories of small states and middle powers. This is puzzling because a number of small states and middle powers, especially in the Asia Pacific and Middle East, have been modernizing their forces by procuring selected advanced military systems and technologies while studying the applicability of selected U.S. RMA-oriented concepts in their military doctrines (Lee 2003). Notwithstanding the varying pace, character, and impact of the RMA diffusion across different geographical regions, organizational structures, and strategic cultures, there are at least three underlying drivers that have accelerated its diffusion: (1) the persisting geopolitical insecurity, regional rivalry, and uncertainty stemming from the emergence of complex types of conflicts and threats of the twenty-first century that have broadened national defense requirements and operational needs; (2) the increased economic growth rates, particularly in East Asia and the Middle East, that have increased the capacity to purchase cutting-edge weapon systems and accelerated force modernization programs; and (3) the rapid technological change embedded in the globalization, consolidation, and competition of global arms markets and defense industries that have led to diversification in their commercial interests through export-oriented strategies and innovation (Bitzinger 2003). Ultimately, however, military organizations are seen both as comparative and competitive institutions that closely monitor each other and calibrate their performance in relation to other militaries. A military capability proven successfully on the battlefield can stimulate responses abroad – to emulate, offset, or innovate (Goldman and Eliason 2003, 5).

Accordingly, the fundamental underpinnings of RMA-oriented military innovation – as a theory, process, and debate – cannot be confined solely to U.S. perspectives alone. The relative dearth of scholarly literature on global RMA diffusion paths and patterns represents the key departure point for this book. In particular, this study attempts to contribute to the scholarly and policy-oriented literature in strategic studies and international security by addressing the broad puzzle of how the global diffusion of RMA-oriented military innovation – the process of international transmission, communication, and interaction of RMA-related military concepts, organizations, and technologies – has shaped the paths, patterns, and scope of military innovation of selected small states. In a reverse mode, how have selected small states influenced the conceptualization and transmission of the RMA theory, processes, and debate? Furthermore, why and when do selected small states decide to pursue an RMA-oriented military modernization? And what are the key theoretical and policy ramifications of RMA diffusion for the security of small states?

Implicit in these questions is the hypothesis that RMA diffusion outside the great power context – its pace, character, drivers, magnitude, and impact – has *reflected varying evolutionary trajectory* shaped by the confluence of three sets of interrelated factors: (1) the level of strategic, organizational, and operational adaptability in responding to shifts in their geostrategic and regional security environment; (2) the ability to identify, anticipate, exploit, and sustain *niche military innovation* – select conceptual, organizational, and technological innovation intended to enhance the military's ability to prepare for, fight, and win wars; and (3) strategic culture. In this context, strategic adaptability refers to the ability to change military posture rapidly and seamlessly amid changes in geostrategic environment or national strategy, while operational adaptability can be conceptualized in the flexibility and robustness of armed forces to employ strategies and tactics in different ways and scenarios (Davis 2000, 31). Strategic culture is a function of the magnitude, character, and impact of "strategic path dependence" – embedded in the continuity and change of prevailing traditional security paradigms that set conditions for the development of strategic and operational thought within armed forces. The ensuing dynamic relationship between strategic adaptability, defense management capacity, and strategic culture results in a much broader variation in the military innovation trajectories of small states. In other words, military innovation and its diffusion in small states is not a linear process, propelled by technological competition among great powers, and then selectively adopted or emulated by less advanced "peripheral" or smaller units. Rather, it can take multiple facets and rarely proceeds in a synchronized rate, path, or pattern. Given the prevailing external and internal variables – enablers and constraints that shape the capacity of states to integrate, adapt, and utilize niche military innovation under local circumstances – the process of RMA-oriented military innovation is not sequential, nor does it follow a particular model. Technological innovation may precede conceptual and organizational adaptation, or conceptual speculation may lead to exploration and experimentation, but not technological implementation. Only if military innovation meets implementation in both policy and strategy, one can theorize about a "radical" or "disruptive" RMA game-changing defense transformation.

Indeed, in a historical perspective, most military innovations have arguably followed a less distinct revolutionary or transformational path, consisting of incremental, often near-continuous improvements in existing capabilities (Ross 2010). In other words, while major, large-scale, and simultaneous military innovation in technologies, organizations, and doctrines have been a rare phenomenon, military organizations have progressed through *a sustained spectrum of innovations* that have shaped the conduct of warfare (Cheung, Mahnken, and Ross 2011, 80). Emily Goldman, for example, captured the spectrum in terms of "adaptive strategic adjustments" or minor innovations to existing goals, strategies, and/or structures, and "innovative strategic adjustments" that define MMIs (Goldman 1998, 233–267). Peter Dombrowski and Andrew Ross (2008) coined the term "modernization plus," in which military innovation reflects rather an evolutionary change characterized as "relevant upgrades or improvements of existing military capabilities through the acquisition of new imported or indigenously developed weapons systems and supporting assets, the incorporation of new doctrines, the creation of new organizational structure, and the institutionalization of new manpower management and combat training regimes" (Tellis and Wills 2006, 15).

Military innovation spectrum must, therefore, also "include not only the actual instruments or artifacts of warfare, but the means by which they are designed, developed, tested, produced, and supplied – as well as the organizational capabilities and processes by which hardware is absorbed and employed" (Ross 1993, 111). In other words, technological innovation provides a necessary but not a sufficient condition in pursuing and implementing the RMA. As Dima Adamsky noted, "RMAs are driven by more than breakthroughs in technology, which in themselves do not guarantee successful innovation" (2010, 1). John Garstka conceptualized the varying spectrum though six specific types of innovation relative to the broader political, socioeconomic, strategic, and cultural conditioning factors shaping the character and conduct of warfare: theory, technology, process, organization, people, and concepts (Garstka 2009, 57–79). *Theoretical innovation* such as theory of aerodynamics, operations research, and energy maneuverability theory provides the underlying knowledge that enables further innovation. *Technological innovation* involves the development of new materials or technological capabilities in both civil and military domains and exhibits concomitant *spin-on or spin-off effects*. Closely linked with technology is *process innovation* that relates to how military organizations adopt and adapt new doctrines, techniques, procedures, and training in their operational conduct. *Organizational innovation* consists of the development of new organizational commands, branches, and units such as special operations commands, space and cyber commands. *People innovation* creates new means for developing, employing, and retaining human capital as well as changing organizational culture – that is, organizational learning, creating new paths. And *concept innovation* describes or articulates new approaches or concepts for combining technology, processes, organization, and people. All six types affect the capacity of states to generate *military effectiveness*, or the "capacity to create military power from state's basic resources in wealth, technology, population size, and human capital" (Garstka 2009, 58). This includes four mutually reinforcing attributes: *integration* of military activities within and across different levels; degree of *responsiveness* to internal constraints and to the external environment; high *skill* in the motivation and basic competencies of personnel; and high *quality* in the weapons platforms, technologies, and equipment. Taken together, the spectrum of military innovation links *strategic innovation* focusing on large-scale changes in grand strategy, defense strategy, and military strategy; *defense innovation* embedded in the transformation of ideas and knowledge into new or improved products, processes, and services for military and dual-use applications; and *military innovation*, encompassing "both product innovation and process innovation, technological, operational, and organizational innovation, whether separately or in combination to enhance the military's ability to prepare for, fight, and win wars" (Cheung, Mahnken, and Ross 2011).

Conceptualizing small states

What is a small state? This question has propelled also a wide range of contending debates in the areas of small state studies and international relations for over four decades (Ingebritsen *et al.* 2006). This is because there are inherent conceptual tensions and difficulties in bracketing and defining the "shades of grey" of small states and their relative capabilities, relevance, policy options, and patterns of behavior in the international system. First, a literature survey identifies two contending categories of definitions of small states: *absolute* and *relational* (Wiberg 1987, 339–363). Underscoring both categories is the neo-realist assumption that the behavior of states in the anarchic international system is mainly determined by the power relations and differentials among them (Handel 1990). The former includes indicators of "size" that serve as the basis for classification such as population, area, GNP, size of military forces, and other numerical indicators. These are then compared or correlated with other variables. The smallness of a state is then a function of comparative quantitative indicators, defined largely in absolute terms. The absolute classification, however, has been perceived as rigid and often contested due to its arbitrary nature. On the other hand, relational definitions are based on the premise that "smallness" means the *lack of influence* in the international environment, or high sensitivity to the environment, and lack of immunity against influences from it, or both. Implicit here is the notion that small states are essentially weak states. According to David Vital (1967), "the smaller the human and material resources of a state, the greater are the difficulties it must remount if it is to maintain any valid political options at all, and, in consequence, the smaller the state, the less viable it is as a genuinely independent member of the international community." Small states, however, *are not* necessarily synonymous with weak states and vice versa. For example, Singapore is often viewed as a small country. This is certainly the case when it is compared to Indonesia; the latter has a huge archipelago and a population of 242 million, compared to the tiny island state of Singapore, with just 5.5 million people. However, Singapore's defense budget is twice that of Indonesia. As Michael Handel noted, "in the study of international relations, it is not the *size* of a state which matters, but rather its *relative strength*" (1990, 10).

The starting point is thus measuring the power of a state *not* against all other countries or a single source but *in relation to its neighbors* and by the degree to which the strength at its disposal matches its national goals and ambitions. Handel (1990) proposes a dynamic, multi-criteria definition of small states based on population, area, economy, military power, interests, and influence in the international system. In doing so, he examines the internal as well as external sources of states' weakness or strengths embedded in the power constellations of different international systems. He warns that it is impossible to define groups of states in the international system with a simple set of tangible measures into "one concise, precise, and elegant statement" as their position is constantly being challenged, tested, and changed. In other words, applying quantitative variables such as population or territorial size, GNP, size of armed forces, and other measures to determine the "smallness" or evaluate states' position in the international system significantly neglects other important *qualitative* or intuitive criteria such as national character or "effective population" that impacts the state's economic, industrial, and social capacity.

The purpose of this book, however, is not to dwell on debating the validity of conceptually precise definitions of "small states" that may be essentially arbitrary. As David Vital (1971) noted, "a loosely defined concept of a small state that eschews rigid specifications is preferable to a precise definition." Rather, it is useful to conceptualize baseline characteristics of *small states that are relevant for the study of military innovation.* Accordingly, this book defines "small states" in the broader context of the following three baseline indicators: (1) *geostrategic constraints,* (2) *relative capabilities,* and (3) *patterns of behavior.* First, small states in this study are conceptualized along Handel's *relational* approach, defined in the prevailing *geostrategic predicaments* in their respective regional or geographic settings and configurations. The common denominator is their prevailing sense of insecurity and drive to ensure their existence and survival in conditions of strategic inferiority. In this context, there are at least five characteristics of "smallness":

1   relative asymmetries of the location, geography, and size;
2   demographic, economic, and natural resource constraints;
3   dependence on external political and material support;
4   security uncertainties or proximity to areas of conflict;
5   relations and importance between and to great powers.

The sum of these factors then shapes their security equation – from the varying strategic assessments and defense policies toward military capabilities and force employment, and their patterns of behaviors in the international arena. In this neorealist theoretical perspective, the anarchic nature of the international system determines the behaviors of states and vice versa. Small states have traditionally not been able to significantly shape or change the system-wide forces that influence them (Keohane 1969, 291–310). The patterns of behaviors by small states in the international system typically include low-level participation in world affairs; narrow scope of foreign policy issues; limitations of behavior to their immediate geographic arena; reliance on superpower for protection, partnerships, and resources; active membership in multinational institutions; disproportionate spending of foreign policy resources to ensure physical and political security and survival (Hey 2003).

Notwithstanding the varying geostrategic predicaments that have traditionally shaped national security and military strategies of small states, not all small states in this definition are pursuing military innovation. Not only the *motives* but also the *means* and *sophistication* are essential criteria for small states pursuing military innovation. By means and sophistication, this study takes into account (1) defense management capacity to plan, organize, lead, and control armed forces and their supporting systems to adopt particular military innovation trajectory while sustaining sufficient military capabilities to achieve national security objectives; (2) relatively advanced, reliable, cost-effective DIB capable of developing innovative defense technologies, niche products, and services; and (3) combat proficiency to engage in a range of military operations and having the potential to integrate and exploit selected innovative concepts and technologies at the operational level.

Throughout the Cold War, the range of viable foreign and defense policy options for small states to deal with their security predicaments was not wide. Not all small states have been necessarily considered weak states; however, they have been more dependent on outside political and material support, particularly by great powers. On the one hand, small states have aimed at strengthening alliances with great powers – a form of *external balancing* (Waltz 1979) – in which a great power defends the interests of a small state in subordinate domains and assures at least partial extended deterrence, but at the same time it may impose costly diplomatic attachments and long-term policy constraints. Alternatively, small states have pursued self-reliance by maximizing their available internal resources, a form of *internal balancing*. However, as Stuart Cohen noted, the potential "pay-offs" of internal balancing has been limited, as small states, especially "those seeking to counterbalance their 'smallness' by increasing levels of military expenditure and production invariably find that the ancillary economic and social costs associated with pursuing self-reliance are always high, and in many cases crippling" (1995, 78–93). Notwithstanding alternative strategies for small states such as "defensive isolation" (*Systemschliessung*), neutrality or diverse strategies of adaptation such as "active foreign policy" or "non-offensive defense," the structure and realities of the anarchic international system have arguably forced most small states to adopt defensive postures based on a mix of both external and internal balancing (Inbar 1996).

With the end of the Cold War era, however, and the resulting systemic shifts in the international arena in the twenty-first century characterized by changes in the balance of power, the increasing salience of non-state actors, sociocultural and economic globalization, and the information revolution, the traditional security equation of small states has been changing. According to Efraim Inbar, while the impact of these shifts varies in different geographical areas of the globe, two key variables in the security paradigm of small states have been affected: *freedom of action* and *relative capabilities* (1997, 1–9). First, the demise of the communist Soviet bloc and the prevalence of the United States as the sole global superpower alleviated the political constraints traditionally imposed by superpowers on small states during the Cold War, permitting relatively greater freedom of action to pursue their strategic interests. In this context, regional actors that have been once constrained by the fear of superpower intervention can pursue their national interests with fewer concerns on the global implications of their policies (Inbar and Sandler 1994, 330–358). At the same time, small states that face or are threatened by aggression can no longer be assured of superpower intervention for their defense, which essentially amplifies their security dilemmas.

Second, the prevailing insecurity and uncertainties of the post-Cold War era has subsequently increased the defense requirements of many small states. In particular, the means and modes of conflict in the twenty-first century have become more dispersed and more decentralized, with the convergence of conventional, low-intensity, asymmetrical, and non-linear threats blurring traditional offense and defense lines. The progressive complexity of emerging threats and conflicts thus means the widening of the scope and character of operational requirements. Modern military organizations are ever more required to field multi-mission capable forces backed by advanced military technologies that are adaptable and interoperable. This in turn amplifies the relevance of RMA-oriented military innovation that emphasizes networked systems, platforms, organizations, and concepts to achieve greater efficiency and effectiveness in the use of force. In this context, the ongoing globalization of arms markets, arms competition, and privatization of defense industries may enable selected small states to gain access to advanced military technologies seen as significantly enhancing their defense capabilities.

Pursuing military innovation by small states, however, is bound to a range of policy impediments, feasibility risks, and challenges – from organizational and institutional barriers to change, to designing new doctrines and operational concepts, to budgetary and resource constraints that limit the options for weapon systems, procurement, and acquisition. According to Treddenick, "managing the

RMA means finding the resources to make it a reality – the essential issue is affordability" (2001, 97). In other words, as Horowitz argues, "as the cost per unit of technological components of a military innovation increases and fewer commercial applications exists, the level of financial intensity required to adapt the innovation increases" (2010, 3). For small states with limited resources, the development, procurement, and integration of RMA-related technologies stipulates even greater need for efficient allocation of scarce defense resources that must be balanced in the context of short- and long-term national security objectives and operational plans – that is, with regard to the existing force structures, quality and quantity of weapons systems available for deployment, operations and maintenance, logistics, infrastructure, and training. In this context, policy makers may face substantial political pressures to downsize defense expenditures in favor of policies that aim at ensuring socioeconomic stability and growth. As a result, the financial constraints coupled with broader resource mobilization challenges have serious implications on the selection, acquisition, and procurement of systems and platforms that characterize the "hardware" side of the RMA.

At the same time, however, enhancing military effectiveness cannot be achieved by simply buying new hardware alone (Isaacson, Layne, and Arquilla 1999). Implementation requires a formal transformation strategy; new units to exploit and counter innovative mission areas; revising doctrine to include new missions; establishing new branches and career paths within the military; changing the curriculum of professional military education institutions; and field-training exercises to practice and refine concepts (Goldman and Mahnken 2004). Defense planners must first define and contextualize the relationship and linkages between the varying RMA concepts at the strategic, operational, and tactical levels. At the strategic level, this means clearly identifying both short-term and long-term strategic risks, resource allocation priorities, and rethinking national security conceptions. At the operational level, defense policy makers must formulate the joint (inter-service) operational concepts, doctrines and training programs, devise specific procurement methodologies for advanced technologies while identifying and solving short-term or current operational problems. The lack of interoperability may create serious problems on the battlefield, with consequences for the political realm (Goldman and Eliason 2003). Furthermore, at the operational level, military innovation may bring new language and terms of reference, changing standing operating procedures, exercise plans, and organizational structures. In short, an RMA-oriented military innovation requires substantial organizational adaptation at virtually all levels. Yet, with limited defense resources, deeply entrenched strategic cultures, organizational resistance to change, and inter-service rivalries and vested organizational interests, the pace and character of RMA implementation may be incremental or gradual.

Notwithstanding the range of defense management problems, perhaps the most compelling impediments to an RMA-style military modernization and innovation for small states is the broader strategic context of the changing global security dynamics and emerging threats. Conflicts in the twenty-first century are characterized by unconventional, unforeseen, and unpredictable methods of warfare. There are a number of low-tech and relatively inexpensive asymmetric responses

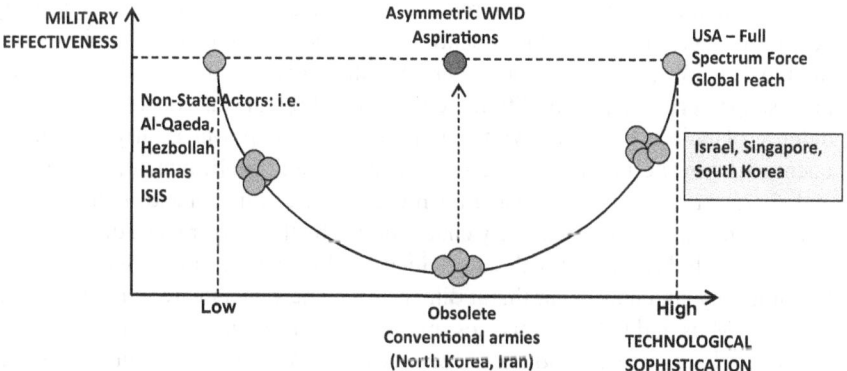

Figure 1.1 Patterns of asymmetric challenge, strategic response, and adaptation

Source: Adapted from Kaspar Stefan Zellweger's Framework, The Geneva Institute of International Studies (2008).

and countermeasures that can selectively mitigate key advantages of RMA-oriented forces. Figure 1.1 illustrates this pattern of asymmetric "challenge, strategic response, and adaptation," where non-state armed groups or terrorist organizations are motivated to achieve a relatively high level of "effectiveness" or political leverage by utilizing a mix of high-tech and low-cost technologies and adapting asymmetric tactical negation not restricted by traditional rules of conflict. They may attempt to subvert the technological advantages of RMA-oriented states through a prolonged, unstructured tactics, seeking to exploit their political and strategic vulnerabilities. Similarly, states with outdated conventional armies that cannot finance or develop modern weapons technologies may opt for asymmetric responses (operational or technical) by pursuing the development and production of weapons of mass destruction (WMDs) and ballistic missiles.

Why study Israel, Singapore, and South Korea's military innovation?

Notwithstanding the conventional wisdom suggesting that Israel, Singapore, and South Korea reflect fundamentally divergent historical experiences, security environments, political and economic development models, varying strategic and military cultures, and thus their RMA trajectories cannot be compared, this study argues that the three states represent relevant cases for a comparative study of RMA diffusion and adaptation (Figure 1.2). To begin with, all three states can be considered small states from the *relational* point of view (Handel 1990) – based on the relative asymmetries of their location, geography and size; demographic, economic, and natural resource constraints; dependence on external political and material support, particularly the United States; security uncertainties or proximity to areas of conflict; and relations and importance between and to great powers.

All three states have historically faced an array of persistent security challenges and uncertainties brought by the realities of their external and internal security conditions, asymmetries of their location, size, and geopolitical constraints. During the Cold War, Israel, Singapore, and South Korea's security environment has been shaped primarily by traditional conventional threats and linear threat-based defense planning. However, with the end of the superpower rivalry and subsequent shifts in the international and regional strategic environment, the sources and characteristics of threats have been changing. In this context, their security environment has been increasingly characterized by the convergence of even more complex or "hybrid" security threats, which combine conventional, asymmetrical, low-intensity, and non-linear threat dimensions. These include regional proliferation of WMDs and ballistic missiles, maritime security, information warfare and cyber-attacks, as well as traditional security threats posed by conventional power projection aspirations and force developments by neighboring states. The increasing amalgamation of security threats has in turn created greater security uncertainties and defense policy challenges, which have propelled robust debates within their strategic and policy communities regarding the relevance of their traditional security paradigms, direction and scope of their particular force modernization programs, defense requirements, and overall strategic choices.

Indeed, facing changing strategic realities, Israel, Singapore, and South Korea have been rethinking their traditional security conceptions, defense planning and management processes, and force postures. While their defense concepts, strategic adaptations, and operational conduct have varied, the select states have essentially searched for new "force multipliers" of their military capabilities. In the process, Israel, Singapore, and South Korea have been studying, benchmarking, and debating selected RMA concepts and technologies, adopting them based on their defense requirements, available resources, and operational lessons learned. In Israel's case, the practical implementation of RMA-oriented tactics by the Israel Defense Forces (IDF) has even preceded both the American and Soviet conceptual formulation. As Dima Adamsky points out, "while the Americans were the first to produce RMA-type munitions, and the Soviets were the first to theorize on their revolutionary implications, the Israel Defense forces were the first to employ these revolutionary weapons in combat" (2010, 4).

Notwithstanding their innate geostrategic conditions of vulnerability coupled with the increasing convergence of emerging asymmetric threats, Israel, Singapore, and South Korea share three additional "similarity factors": (1) *defense management capacity* to cope with the ever-changing strategic and operational changes, resource constraints, while simultaneously developing national innovation strategies to sustain military-technological edge within acceptable costs and risk framework; (2) small but advanced DIB capable of research, development, testing, and evaluation of selected weapon technologies, niche products, and services based on their armed forces' needs. In particular, Israel, Singapore, and South Korea's DIB share similarities in their indigenous production of advanced military systems and technical expertise to upgrade imported systems to meet specific requirements. Their defense industries, to a varying degree, are able to adapt to global commercial arms markets by fostering high-technology defense–industrial capacity, innovation, and competitiveness through a defense-related R&D–civil research interaction, prioritization, and diversification of defense production. This includes specialization in upgrading or modifying major weapons systems as well as developing "niche" weapon component parts and sub-systems, sub-assemblies, and components available for exports – from complete systems at the high end to low-level components at the bottom. Finally, Israel, Singapore, and South Korea's armed forces would be characterized as well-trained forces with a sufficient *combat proficiency* to engage in divergent sets of military operations and, in doing so, have the potential to integrate and exploit selected RMA concepts, organizational, and technological innovations. Unlike superpowers, resource-constrained small states such as Israel, Singapore, and South Korea are realistically not able to achieve high level combat proficiencies in *every* operational mode; however, their armed forces should be adequately trained and prepared to operate in select areas of high operational complexity – that is, network-centric C4ISR, precision strike capabilities, and platforms that provide limited power projection capabilities in air, ground, naval, and cyber domains.

Overall, all three states have experienced a number of challenges in their search for military innovation – from defining new strategic and operational concepts through a broad range of defense management constraints and, perhaps most importantly, by the emergence of new asymmetric forms of warfare. Indeed, the task of formulating a comprehensive strategic blueprint of what constitutes an Israeli, Singaporean, or South Korean RMA has proven challenging. Security planners in all three states must answer anew to what degree are their traditional defense planning paradigms relevant in meeting their continuing conventional security threats as well as a wide spectrum of emerging conflicts and non-linear crises? How to resolve the gap between long-term strategic planning and short-term operational requirements? What weapons technologies and systems should be procured at what price, and which of them are relevant within an affordable framework? What elements of the RMA are relevant in the broader strategic context of the twenty-first century? Accordingly, Israel, Singapore, and South Korea may represent relevant empirical cases not only for testing the validity of the RMA hypotheses but also from the policy-oriented perspective on how to cope with security and defense management challenges posed by the processes of military innovation.

In drawing close parallels between South Korean, Singaporean, and Israeli security challenges, there are words of caution. Several important points should be noted. First, their regional security environment and strategic realities do not reflect an exact microscopic copy of one another. Indeed, many Israeli, Singaporean, and South Korean defense experts would argue that the conflict spectrum of each state remains fundamentally unique; with Israel traditionally facing a dynamic, active, and more intensive conflict spectrum, whereas South Korea's threat environment reflecting rather static defensive patterns (despite North Korean provocations). Moreover, since the end of the Korean War, and for over 60 years onward, South Korea defined its security through collective defense mechanisms institutionalized

	ISRAEL	SOUTH KOREA	SINGAPORE
Geostrategic Predicaments (Baseline)	– Lack of strategic depth; – Resource asymmetry constraints; – Demographic asymmetries; – Hybrid conflict spectrum: surrounded by state/non-state adversaries; – Profound strategic interests and involvement of great powers;	– Profound strategic interests and involvement of great powers; – Hybrid conflict spectrum: North Korea; – Small and highly populated combat radius; – Resource asymmetry constraints; – Demographic constraints;	– Geographical and strategic predicaments – location, size, lack of natural resources, and physical limitations; – Asymmetries in demographic factors; – Profound strategic interests and involvement of major powers; – Historical experience of vulnerability;
Defense Industrial Base	– Small but advanced defense industry: – High-tech arms exports;	– Adapters and modifiers of advanced technologies; – High-tech civil-military industries;	– Advanced "defense ecosystem" – integration of the users (SAF), developers, and producers (local defense industries)
Defense Management Capacity	– Selective acquisition & procurement strategies; – RMA-based military modernization programs;	– Selective acquisition & procurement strategies; – RMA-based military modernization programs;	– Selective acquisition & procurement strategies; – RMA-based military modernization programs;
Combat Proficiency	– Extensive & frequent combat experience;	– Periodic combat experience; – U.S.-ROK training/combined forces;	– Technologically-advanced, operationally-ready force; – Overseas training and participation in operations other than war;

Figure 1.2 Why study Israel, Singapore, and South Korea's military innovation paths?

in the U.S.–ROK alliance, which provided assurances of defense and deterrence vis-à-vis external military threats. On the other hand, Israel's national security conceptions, defense strategies, and operational conduct have been traditionally focused on pursuing self-reliance, albeit recognizing the impediments of super-power involvement coupled with external dependence on major weapons systems procurement. This dimension has shaped the varying motivations for adopting RMA-related concepts and technologies. Second, their internal political and social fabric along with strategic culture and military thinking has varied. Third, the size of their forces, organizational force structure, and weapons systems deployment has also differed, particularly in terms of quality and quantity. These claims, while valid, do not detract from the validity of exploring their search for military innovation beyond the differences in circumstances.

Chapter themes and overview

This book applies a comparative case study method to investigate RMA-oriented diffusion paths and patterns of military innovation in selected small states – Israel, Singapore, and South Korea – since the mid-1990s. According to Charles Ragin, case studies are used inductively to develop and refine typological theory through a "building block" approach, which is particularly useful in new or emerging research programs to generate theory (Ragin 1994). Scholars such as Alexander George, Robert K. Yin, Bruce Berg, and Stephen Van Evera argue that case studies are well suited for theory testing and the application of theories to explain specific cases and are relevant in exploring *how* and *why (or why not)* types of questions that deal with contemporary phenomena within a real-life context (Yin 1994; Stephen 1997; Berg 2007). Notwithstanding the recurrent trade-offs and inherent limits of case studies, including problems with case selection, the trade-off between parsimony and richness, and the related tensions between achieving high internal validity of particular cases versus making generalizations applicable to broad populations (George and Bennett 2005, 22), a comparative case study method may provide insights underlying the processes of RMA-oriented military innovation in selected small states. Underscoring this book are two key questions related to the intellectual history, processes, and debates that have shaped military innovation trajectories in the select states. First, to what level have the select small states acknowledged RMA-oriented conceptual, organizational, and technological military innovation as a disruptive paradigm shift or as a "new theory of war" in their defense strategies, military doctrines, and operational concepts? Second, how have these states interpreted and defined the relevance, character, and applicability of the RMA in their strategic and military narratives? In exploring these two questions, the first step is to examine the continuity and change in the security conceptions, defense strategies, and military doctrines of the select states, particularly reflecting on the changes in the sources and character of security threats and "tipping points" in their strategic assessments. Within this context, one has to identify the introduction of specific RMA-related conceptual elements and schools of thought, tracing their historical development – from the intellectual roots to more sophisticated phases of adaptation, modification, and potential innovation. The resulting case study description then provides essential background and evidence for assessing how, why, and when selected small states conceptualized the RMA in their military innovation paths and patterns.

This book draws on multiple data sources consisting of primary sources based on semi-structured expert interviews, official documents, and a range of secondary literature and published sources that trace military innovation concepts, processes, and outcomes in Israel, Singapore, and South Korea over the past three decades. According to Robert Yin, "a major strength of case study data collection is the opportunity to use many different sources of evidence," which enables addressing a broader range of historical, attitudinal, and behavioral issues as well as exposes converging lines of inquiry and ultimately enables continuous interaction between the theoretical or policy issues being studied and data being collected (Yin 1994, 97). A multiple source framework is thus essential not only to strengthen the construct validity and reliability of a multiple case study design but also to overcome the inherent difficulties in obtaining data on sensitive issues relating to Israel, Singapore, and South Korea's security. The continuing secrecy that governs the norms, communication, and access to military sources in the three countries is prevalent and perceived as necessary in conditions of constant security risks. Israel, for example, does not publish any official data on order of battle, force structure, or doctrine. IDF officers below the rank of brigadier general appear in the press or in professional publications identified only by their first name, or even their first initial, and Israeli military archives are not open to researchers. Interviews with active IDF officers must be approved by the IDF Spokesperson's Office. Similarly, South Korea's Military Secrecy Protection Law sets the rules on disclosure of information pertinent to national security. To mitigate the above constraints, this book combines both primary and secondary sources: available published materials and documents on national security and defense policy issues in Israel, Singapore, and South Korea, drawn from leading research institutes, newspapers, books, journals, and government publications. In Israel, for example, these include articles, papers, books, and journals published by the IDF – *MaArachot*; Institute for National Security Studies based in Tel Aviv, Begin-Sadat Center for Strategic Studies at Bar-Ilan University, Fisher Brothers Institute for Air and Space Strategic Studies in Herziliya, Jerusalem Center for Public Affairs, as well as newspaper articles published in major Israeli newspapers such as *Ha'aretz*, *Yedioth Ahronoth*, and *Jerusalem Post*. At the same time, the book relies on intensive on-site expert interviews conducted with a number of Israeli, Singaporean, and South Korean current and former government officials, military officers, academics, policy analysts, and journalists. The interviews have been designed to fill the gaps in the literature and provide key insights into the fundamental issues, drivers, and limitations facing Israeli, Singaporean, and South Korean military decision making, their views on the role of technology in warfare, perceptions of RMA and future battlefield, problems and challenges of defense management and military modernization, and lessons learned in the use of force. The primary interview sites in Israel included IDF Dado Center for Advanced Military Studies, Institute for National

Security Studies, Begin-Sadat Center for Strategic Studies, Truman Research Institute at Hebrew University of Jerusalem, Harold Hartog School of Policy and Government at Tel Aviv University, Interdisciplinary Center in Herzliya, as well as a number of individual interviews with former high-ranking government and intelligence officials who served in the National Security Council, Mossad, and the IDF, most of whom have requested anonymity. In South Korea, the main interview sites included the Ministry of National Defense, Korea Military Academy, Korea Institute for Defense Analyses, Yonsei University, as well as interviews with former high-ranking military officers in the ROK Army and Air Force, and members of the Presidential Policy Advisory Committees on National Security. In Singapore, the author interviewed academics, policy analysts, former Singapore Armed Forces (SAF) officers based at the S. Rajaratnam School of International Studies, Goh Keng Swee Command and Staff College, the Lee Kuan Yew School of Public Policy, and the Future Systems and Technology Directorate at the Ministry of Defense, coupled with secondary literature in journals, such as the *Pointer – Journal of the Singapore Armed Forces*.

Taken together, the book is organized as follows: Chapter 2 provides a comprehensive literature overview of the general RMA debate, its intellectual history, diffusion, and schools of thought that have emerged over the past two decades. It sketches the evolving character and direction of the RMA conceptual debate on both theoretical and policy-oriented levels and argues that issues surrounding the RMA and its diffusion, emulation, and adaptation in the security, defense planning, and use of force by selected small states provide relatively unexplored territory in the ongoing RMA discourse. To demonstrate this, the chapter provides an overview of the stages or "Six RMA Waves" through which the RMA debate has evolved – primarily but not exclusively in the U.S. military thought. In doing so, the chapter attempts to contextualize the varying approaches in understanding the RMA and emphasizes that the vast majority of RMA and defense transformation studies have been relatively silent or ignored the implications of RMA diffusion and adaptation in security, defense planning, and the use of force by small states and middle powers, particularly in the strategic contexts of East Asia and the Middle East.

Presenting the first case study, Chapter 3 explores the conceptual diffusion, adoption, and adaptation of the RMA in Israel – through the lens of continuity and change of its national security conceptions, defense strategies, and operational experience of IDF. The chapter argues that development of Israeli RMA concepts from the mid-1990s onward should be conceptualized primarily through the lens of continuity and change in Israel's security conceptions, focused on retaining Israel's qualitative "strategic edge" amid continuously changing operational requirements. In this context, Israel has been one of the first countries to apply RMA-related technologies in combat in the early 1980s under the conceptual umbrella of "integrated battle." IDF's experiences have inherently influenced the American, Soviet, and European strategic perspectives and debates on the future of warfare. However, until the late 1990s, the IDF has not viewed the emergence of the RMA as a relevant paradigm shift, nor has initiated a comprehensive and disruptive defense transformation drive. Rather, Israel's RMA discourse has reflected a continuous debate between the proponents of traditional concepts and reformers – those arguing for new military thinking within the IDF. Consequently, the changing strategic realities and operational experience over the last two decades forced the IDF to rethink its traditional threat-oriented concepts and experiment with innovative combat tactics at different levels of warfare. The culmination of Israeli RMA-oriented conceptual military innovation and debate has reached in the mid-2000s, when the IDF experimented with a theory of Systemic Operational Design and its integration into a new Concept of Operations (CONOP), which has been subsequently questioned, abandoned, modified, and revised. In short, Israel's military innovation has been driven by operational and tactical application – developments oriented as a response to specific threats

Chapter 4 projects South Korea's RMA trajectory through the complexity of shifts in the security dynamics on the Korean peninsula. Since the early 1990s, South Korea has been attempting to undertake a comprehensive military modernization in order to respond to the widening spectrum of threats, mitigate technological and interoperability gaps with the U.S. forces, and eventually attain self-reliant defense posture. In the process, South Korean RMA diffusion path proceeded on two interrelated levels: external/emulative, shaped by the shifts in the U.S.–ROK alliance and U.S. defense transformation; internal/adaptive, defined in South Korea's changing strategic assessments and policy imperatives to develop "self-reliant" military capabilities. However, the compelling and relatively ambitious character of Korean RMA-oriented defense plans has been in sharp contrast to the prevailing structural and political realities such as contrasting calibrations of defense requirements, structural dependence on the U.S.–ROK alliance, static, defensive force posture, and asymmetric organizational force structure that have sustained the relevance of traditional security concepts. As the chapter argues, there has not been a distinct Korean RMA-oriented conceptual innovation toward a new theory of war, rather progressive shifts from operational and military-technological emulation to selective capability adaptation in the gradual/evolutionary process of military modernization.

Chapter 5 attempts to project the essential contours of the continuity and change in Singapore's security paradigm and defense strategy; the intellectual history and processes that have shaped the direction, pace, and character of military modernization in the SAF; and concepts of future warfare presently emerging in Singapore's strategic thought. In doing so, the chapter argues that the character of Singapore's force modernization has projected a continuous evolutionary path in terms of its systems and structures – a structured-phased development: the first-generation SAF (1960s–1970s) focused on capability development of individual services; the second generation reflected a period of consolidation and adaptation from service-oriented strategic thinking toward conventionally oriented combined arms warfare (1980s–1990s), and the third-generation SAF (2000 onwards) has aimed at implementing a transition toward a joint strategy for multi-mission type forces with capabilities ranging from defense diplomacy to select kinetic integrated strike capabilities against a wide spectrum of threats. In the process, SAF's doctrinal orientation and operational conduct has shifted significantly in its character: from a purely island defensive "poisoned-shrimp" strategy in the 1970s, which envisioned high-intensity urban combat to impose unacceptable human and material costs to potential aggressors, toward a "porcupine" strategy of the 1980s that developed a limited power projection in Singapore's near seas and envisioned a preemptive posture by transferring a potential conflict into enemy's territory, to the ongoing 3G concepts in the 2010s analogous to a "dolphin" strategy – a "smart" or networked SAF leveraging not only precision fires, maneuver, and information-superiority capabilities but also operations other than war in geographically distant areas from Singapore. In the process of pursuing military innovation, however, the SAF has inherently faced a number of challenges: at the strategic level, the question of how to strike a balance between preserving tried-and-tested strategies and force structures (based on a traditional conscription model) with implementing advanced operational concepts, organizational structures, and technologies relevant for the type of future "hybrid conflicts." At the operational level, the SAF has demonstrated greater technological dependencies that have, on one hand, propelled excellent "joint" tactical experimentation but, at the same time, invoked growing perceptions of "technological superiority" as the primary determinant, enabler and catalyst, of military effectiveness.

Notwithstanding these barriers, the chapter shows that trajectory of military innovation in Singapore transcends purely military-technological and operational domains – it must be situated in the broader context of strategic nexus of Singapore's external foreign/defense and internal public policies that shaped Singapore's readiness and responses to the gradually changing global and regional security challenges. The principal strategy underlying Singapore's security has been the concept of "Total Defense" – a form of national security strategy that has for over 30 years aimed at strengthening and mobilizing resources in five mutually supportive defense domains: military, civil, economic, social, and psychological. In this context, Singapore has been able to offset an array of potential security predicaments through a *comprehensive defense framework* embedded in civil–military strategic interactions at various levels. These include the symbiotic relationships between Singapore's defense spending, economic development, educational system, civil service, media information sphere, and the public. However, with the increasing regional tensions and strategic uncertainties coupled with gradual yet profound changes in Singapore's internal political and socioeconomic fabric over the past five years, the effectiveness of Singapore's "Total Defense" strategies is increasingly challenged, particularly in terms of managing and responding to potentially more severe, cascading, multi-level crises – whether internal or external.

The analytical heart of the book is in the comparative analysis of Chapter 6 that provides a net assessment of Israeli, South Korean, and Singaporean military innovation trajectories and their variations. The chapter applies the methodology of "structured focused comparison," embedded in two main frameworks of analysis: (1) "Modified RMA Diffusion Diagnostics" model that that synthesizes three levels of analysis: *Motives and Models*; *Paths and Patterns*; *Enablers and Barriers of Military Innovation* based on frameworks by Emily Goldman and Thomas

Mahnken (2004, 1–21). (2) The study then conceptualizes a "RMA Diffusion Dynamics" model that enables to locate and compare the varying military innovation trajectories by triangulating *Paths* – emulation, adaptation, and innovation; *Patterns* – speculation, experimentation, and implementation; *Magnitude* – exploration, modernization, and transformation. In this context, the chapter argues that Israel's military innovation trajectory over the last two decades reflects a unique pattern of early adoption/implementation, experimentation, and speculation in the context of multiple operational adaptations, which subsequently led to technological/tactical military innovation. South Korea's trajectory of emulation and adaptation emanates a significant U.S. imprint, shaped by the changes in the U.S. defense strategies, operational concepts, and force postures, as well as South Korea's efforts to minimize prevailing U.S.–ROK interoperability gaps, respond to widening security challenges, and simultaneously develop more "self-reliant" military capabilities. Singapore's evolutionary paths and patterns of military innovation project a classical trajectory of speculation, experimentation, and implementation in the context of ongoing structured-phased capability development. Consequently, the chapters argue that the variance has resulted from the confluence of three sets of variables: (1) the level of strategic, organizational, and operational adaptability in responding to shifts in the geostrategic and regional security environment; (2) the ability to identify, anticipate, exploit, and sustain *niche* military innovation – select conceptual, organizational, and technological innovation intended to enhance the military's ability to prepare for, fight, and win wars; and (3) strategic culture.

The book ends with concluding thoughts on both the theoretical and policy-oriented implications. It shows that the empirical cases of Israel, Singapore, and South Korea inherently challenge traditional "hierarchical" schools of thought explaining military innovation and its diffusion in linear perspectives. Notwithstanding their different paths and patterns, Israel, Singapore, and South Korea's RMA military modernization trajectories reflect an evolutionary rather than revolutionary process of change over the past two decades. While at the strategic level, the military-technical dimension of their RMA-oriented military modernization has not significantly altered or solved their protracted political and security dilemmas, it has nevertheless significantly increased their military capacity to defend themselves against a variety of conventional, low-intensity, asymmetric, and nonlinear threats. In this context, however, the principal challenge for defense planners of small states continues to reflect the same fundamental problem: how to translate limited resources into an effective defense capability amid continuously evolving security challenges?

References

Adams, Thomas K. 2008. *The Army After Next: The First Postindustrial Army.* Westport, CT: Praeger Security International.

Adamsky, Dima. 2010. *The Culture of Military Innovation: The Impact of Cultural Factors on the Revolution in Military Affairs in Russia, the US, and Israel.* Palo Alto, CA: Stanford University Press.

Alberts, David. 1996. *The Unintended Consequence of Information Age Technologies.* Washington, DC: NDU Press.

Alberts, David, John Garstka, and Frederick Stein. 2000 [1999]. *Network Centric Warfare: Developing and Leveraging Information Superiority.* 2nd edition (Revised). Washington, DC: DoD C4ISR Cooperative Research Program.

Berg, Bruce. 2007. *Qualitative Research Methods for the Social Sciences.* Boston, MA: Pearson/Allyn & Bacon.

Bitzinger, Richard. 2003. *Towards a Brave New Arms Industry?* Oxford, UK: Oxford University Press.

———. 2008. "The Revolution in Military Affairs and the Global Defense Industry: Reactions and Interactions." *Security Challenges* 4 (4): 1–12.

Cheung, Tai Ming, Thomas Mahnken, and Andrew Ross. 2011. "Frameworks for Analyzing Chinese Defense and Military Innovation." In *New Perspectives on Assessing the Chinese Defense Economy: 2011 Industry Overview and Policy Briefs*, edited by Tai Ming Cheung, 77–80. San Diego, CA: University of California Institute on Global Conflict and Cooperation.

Clausewitz, Carl von. 1984. *On War.* Translated and edited by Michael Eliot Howard and Peter Paret. Princeton, NJ: Princeton University Press.

Cohen, Eliot. 1996. "A Revolution in Warfare." *Foreign Affairs* 75 (2): 37–54.

Cohen, Stuart. 1995. "Small States and Their Armies: Re-structuring the Militia Format of the Israel Defense Force." *Journal of Strategic Studies* 18 (4): 78–93.

Cote, Owen. 1996. *The Politics of Innovative Military Doctrine: The U.S. Navy and Fleet Ballistic Missiles.* PhD Dissertation. Cambridge, MA: MIT Press.

Davidson, Janine. 2011. *Lifting the Fog of Peace: How Americans Learned to Fight Modern War.* Ann Arbor, MI: University of Michigan Press.

Davis, Paul. 2000. "Defense Planning in an Era of Uncertainty: East Asian Issues." In *Emerging Threats, Force Structures, and the Role of Air Power in Korea*, edited by Chung-in Moon and Natalie Crawford, 25–47. Santa Monica, CA: RAND.

Davis, Vincent. 1967. *The Politics of Innovation: Patterns in Navy Cases.* Denver, CO: University of Denver Press.

Dombrowski, Peter, and Andrew Ross. 2008. "The Revolution in Military Affairs, Transformation and the Defense Industry." *Security Challenges* 4 (4): 13–38.

Eisenstadt, Michael, and Kenneth Pollack. 2001. "Armies of Snow and Armies of Sand: The Impact of Soviet Military Doctrine on Arab Militaries." *Middle East Journal* 55 (4): 905–34.

Farrell, Theo. 1998. "Culture and Military Power." *Review of International Studies* 24: 407–16.

———. 2008. "The Dynamics of British Military Transformation." *International Affairs* 84 (4): 777–807.

Farrell, Theo, and Terry Terriff. 2002. *The Sources of Military Change: Culture, Politics, Technology.* Boulder, CO: Lynne Rienner Publishers.

Farrell, Theo, Terry Terriff, and Frans Osinga. 2010. *A Transformation Gap: American Innovations and European Military Change.* Palo Alto, CA: Stanford University Press.

FitzSimonds, James, and Jan Van Tol. 1994. "Revolutions in Military Affairs." *Joint Forces Quarterly* 4 (Spring): 24–31.

Garstka, John. 2009. "Patterns in Innovation." In *Transforming Defense Capabilities: New Approaches for International Security*, edited by Scott Jasper, 57–79. Boulder, CO: Lynne Rienner Publishers.

George, Alexander, and Andrew Bennett. 2005. *Case Studies and Theory Development in the Social Sciences.* Boston, MA: MIT Press.

Goldman, Emily. 1998. "Mission Possible: Organizational Learning in Peacetime." In *The Politics of Strategic Adjustment: Ideas, Institutions, and Interests*, edited by Peter Trubowitz, Emily Goldman, and Edward Rhodes, 233–67. New York, NY: Columbia University Press.

Goldman, Emily, and Leslie Eliason. 2003. *The Diffusion of Military Technology and Ideas*. Stanford, CA: Stanford University Press.

Goldman, Emily, and Thomas Mahnken. 2004. *The Information Revolution in Military Affairs in Asia*. New York, NY: Palgrave Macmillan.

Gray, Colin. 2002. *Strategy for Chaos: Revolutions in Military Affairs and the Evidence of History*. London, UK: Frank Cass.

Grissom, Adam. 2006. "The Future of Military Innovation Studies." *The Journal of Strategic Studies* 29 (5): 905–34.

Halperin, Morton. 1974. *Bureaucratic Politics and Foreign Policy*. Washington, DC: Brookings Institution Press.

Handel, Michael. 1990. *Weak States in the International System*. London, UK: Frank Cass.

Hayes, Bradd, and Douglas Smith. 1994. *The Politics of Naval Innovation*. Newport, RI: Naval War College Press.

Hey, Jeanne. 2003. *Small States in World Politics: Explaining Foreign Policy Behavior*. Boulder, CO: Lynne Rienner Publishers.

Horowitz, Michael. 2010. *The Diffusion of Military Power: Causes and Consequences for International Politics*. Princeton, NJ: Princeton University Press.

Hundley, Richard. 1999. *Past Revolutions, Future Transformations: What Can the History of Revolutions in Military Affairs Tell Us About Transforming the U.S. Military?* Santa Monica, CA: RAND.

Inbar, Efraim. 1996. "Contours of Israeli New Strategic Thinking." *Political Science Quarterly* 111 (1): 41–64.

——. 1997. "Introduction." In *The National Security of Small States in a Changing World*, edited by Efraim Inbar and Gabriel Sheffer, 1–9. London, UK: Frank Cass.

Inbar, Efraim, and Shmuel Sandler. 1994. "Israel's Deterrence Strategy Revisited." *Security Studies* 3 (2): 330–58.

Ingebritsen, Christine, Iver Neumann, Sieglinde Gstohl, and Jessica Beyer. 2006. *Small States in International Relations*. Seattle, WA: University of Washington Press.

Isaacson, Jeffrey, Christopher Layne, and John Arquilla. 1999. *Predicting Military Innovation*. Santa Monica, CA: RAND.

Keohane, Robert. 1969. "Lilliputians' Dilemmas: Small States in International Politics." *International Organization* 23 (2): 291–310.

Kier, Elizabeth. 1997. *Imagining War: French and British Military Doctrine Between the Wars*. Princeton, NJ: Princeton University Press.

Krepinevich, Andrew F. 1994. "Cavalry to Computer: The Pattern of Military Revolutions." *The National Interest* 37 (Fall): 30–43.

Lee, Chung Min. 2003. "East Asia's Awakening from Strategic Hibernation and the Role of Air Power." *Korean Journal of Defense Analysis* 15 (1): 219–74.

Loo, Bernard. 2009. *Military Transformation and Strategy: Revolutions in Military Affairs and Small States*. New York, NY: Routledge.

Mahnken, Thomas, and James FitzSimonds. 2003. *The Limits of Transformation: Officer Attitudes toward the Revolution in Military Affairs*. Newport, RI: Naval War College Press.

Mazarr, Michael. 1994. *The Revolution in Military Affairs: A Framework for Defense Planning*. Carlisle, PA: Strategic Studies Institute.

Murray, Williamson. 1999. "Does Military Culture Matter?" *Orbis* 45 (1): 27–57.

Murray, Williamson, and MacGregor Knox. 2001. *The Dynamics of Military Revolution (1300–2050)*. Cambridge, UK: Cambridge University Press.

Nagl, John. 2002. *Counterinsurgency Lessons from Malaya and Vietnam*. Westport, CT: Praeger.

Pierce, Terry. 2004. *Warfighting and Disruptive Technologies: Disguising Innovation*. London, UK: Frank Cass.

Posen, Barry. 1984. *The Sources of Military Doctrine: France, Britain, and Germany Between the World Wars*. Ithaca, NY: Cornell University Press.

Ragin, Charles. 1994. *Constructing Social Research: The Unity and Diversity of Method*. London, UK: Pine Forge Press.

Rosen, Stephen Peter. 1988. "New Ways of War: Understanding Military Innovation." *International Security* 13 (1): 134–68.

———. 1991. *Winning the Next War: Innovation and the Modern Military*. Ithaca, NY: Cornell University Press.

Ross, Andrew. 1993. "The Dynamics of Military Technology." In *Building a New Global Order: Emerging Trends in International Security*, edited by David Dewitt, David Haglund, and John Kirtland, 106–40. Oxford, UK: Oxford University Press.

———. 2010. "On Military Innovation: Toward an Analytical Framework." *IGCC* (Policy Brief No. 1): 1–4.

Shimko, Keith. 2010. *The Iraq Qrs and America's Military Revolution*. Cambridge, UK: Cambridge University Press.

Stephen, Van Evera. 1997. *Guide to Methods for Students of Political Science*. Ithaca, NY: Cornell University Press.

Tellis, Ashley, and Michael Wills. 2006. *Strategic Asia 2005–06: Military Modernization in an Era of Uncertainty*. Seattle, WA: The National Bureau of Asian Research.

Toffler, Alvin, and Heidi Toffler. 1993. *War and Anti-War: Survival at the Dawn of the 21st Century*. Boston, MA: Little, Brown and Company.

Treddenick, John. 2001. "Financing the RMA." In *Managing the Revolution in Military Affairs*, edited by Ron Matthews and John Treddenick, 97. New York, NY: Palgrave.

Vital, David. 1967. *The Inequality of States: A Study of Small Power in International Relations*. Oxford, UK: Clarendon Press.

———. 1971. *The Survival of Small States: Studies in Small Power/Great Power Conflict*. Oxford, UK: Oxford University Press.

Waltz, Kenneth. 1979. *Theory of International Politics*. Boston, MA: McGraw-Hill Press.

Wiberg, Hakan. 1987. "The Security of Small Nations: Challenges and Defenses." *Journal of Peace Research* 24 (4): 339–63.

Yin, Robert. 1994. *Case Study Research: Design and Methods*. Thousand Oaks, CA: Sage.

2 The "six waves" of RMA theory, process, and debate

Notwithstanding the seemingly perennial body of literature covering the debate on military innovation through the lens of the RMA debate over the last two decades, the vast majority of writings have been silent or ignored the implications of RMA diffusion on military innovation trajectories of small states. Indeed, the persisting thrust in exploring the RMA has reflected predominantly a U.S.-centered discourse that has evolved in concert with shifts in the U.S. military strategy and use of force. With the predominant focus on the American case, one could argue that there is a significant deficit in the existing literature, particularly in the doctrinal, organizational, and technological dynamics surrounding the RMA-oriented diffusion and adaptation in divergent geostrategic settings and environments. In this context, for example, studies on the Israeli, Singaporean, or South Korean perspectives on military innovation, including concepts, processes, and problems, remain for the most part rare. Indeed, there is no systematic comparative study to date that attempts to link these cases and portray the diffusion and impact of the RMA on their military innovation paths and patterns. This is also puzzling, because Israel, Singapore, and South Korea have been studying and debating RMA-related ideas for more than a decade. Indeed, Israel has been the first country that has extensively utilized select RMA technologies in combat prior to reforming its operational concepts, force structure, and broader strategic doctrine (Adamsky 2010, 4). Meanwhile, Singapore has pursued networking of its "force multiplier" capabilities, applicable in areas such as C4ISR and battlefield management, intelligence, precision strike, unmanned systems, as well as cyber warfare. And South Korea has also adopted and adapted select U.S. RMA concepts, primarily to sustain its interoperability requirements of the U.S.–ROK alliance. Accordingly, this book seeks to bridge this gap and connect seemingly contrasting cases in a comparative study of RMA-oriented diffusion paths and patterns of military innovation. In doing so, the study focuses on the theoretical and conceptual dimensions of RMA diffusion – the "software side." The rationale for this is two-fold: (1) technological innovation provides a necessary but not a sufficient condition in pursuing and implementing the RMA. As Dima Adamsky noted, "RMAs are driven by more than breakthroughs in technology, which in themselves do not guarantee successful innovation" (2010, 1). Moreover, (2) RMA-oriented conceptual adaptation may shape, redefine, or reconfigure defense management processes and its interrelated components: technologies, doctrines, organizations.

To begin with, however, it is necessary to establish a baseline for understanding the contemporary perspectives on military innovation through the lens of the ongoing RMA debate – its terminology, historical and conceptual diffusion trajectory, contending viewpoints, as well as its policy-making relevance. While most theorists and policy experts agree that the RMA debate will likely have profound implications on the future modes and means of warfare by shaping the core competencies of military organizations, no clear consensus has emerged on its validity, meaning, magnitude of impact, and direction (Adams 2008, 12–33). A closer look at the RMA literature suggests there is no single "RMA" school of thought; its conceptual development has been subjected to often ambiguous and diverse interpretations, modifications, and contending theoretical and policy debates since its inception. Indeed, there is a range of definitions that attempt to conceptualize the RMA in the context of its *scope, magnitude,* and *speed of change* in the character and conduct of warfare. The theoretical and methodological approaches to the study of the RMA have been further shaped by the varying epistemological traditions and interdisciplinary perspectives based in political science, public policy, defense economics, and strategic studies. In this context, this chapter attempts to construct a historical narrative of the RMA debate – its theoretical origins and policy-oriented diffusion. While by no means complete, the review sketches how the RMA evolved, mapping its conceptual thrust in line with policy adaptation, from its intellectual discovery in the Soviet Union to its adaptation in the United States. In doing so, the review provides a framework for synthesizing the relevant RMA schools of thought that have emerged over time with regard to the (1) definition, metrics, and impact; (2) sources, levels, causes and effects; (3) diffusion, emulation, and adaptation. The review also shows the asymmetries in the existing RMA discourse primarily, although not exclusively, from the standpoint of RMA diffusion and adaptation by select small states.

Conceptual diffusion of the RMA: "six waves"

The conceptual development and arguments underlying the RMA debate have progressed on two parallel levels: *theoretical* and *policy-oriented* (Shimko 2010, 1–25). The former includes empirical studies and historical analyses on the change and continuity in warfare in general, which transcend immediate national concerns. The latter includes policy prescriptive studies and strategic analyses that guide policy makers on a broad range of issues, predominantly focusing on defense transformation imperatives and its challenges. In order to synthesize both theoretical and policy dimensions of the RMA literature, this chapter applies two heuristic devices: (1) a chronological assessment of the stages, constitutive elements, and dominant characteristics that have defined particular RMA phases; and (2) contending schools of thought and their interpretations in the resulting RMA debates.

In particular, the trajectory of the RMA debate can be first projected through a critical *assessment of the stages* though which the RMA theory, process, and narrative

	1980's	1990–94	1995–2000	2001–2005	2005–xxx
Wave	Intellectual Discovery	Early Adaptation in the West	RMA 'Technophilia'	Shift to Defense Transformation	Second & Third Thoughts
Concept	*Soviet Military- Technical Revolution*	*Military Revolutions vs. RMAs*	*Revolution in Military Affairs*	*Defense Transformation*	*Modernization "Plus"*
Focus	Technological paradigm shift; Soviet doctrinal innovation;	RMAs in history; Sources of mil. Innovation;	System of Systems; Network Centric Warfare	Effects-Based Operations; Network-Centric Warfare	From RMA paradigm shift to a "shift in emphasis"
Debate	Is there RMA? What is RMA? When do RMAs happen?		What is RMA? Why RMA?	Is the RMA feasible, affordable, and desirable?	

Figure 2.1 Overview: five RMA "waves"

evolved over the past 20 years. In 2006, Colin Gray briefly mentioned five progressive stages or "RMA waves" (2006, 113–115): (1) the initial intellectual discovery by the Soviet military thinkers in the early 1980s, (2) the conceptual adaptation, modification, and integration in Western strategic thought during the early 1990s, (3) the climax of the RMA debate during the mid-to-late 1990s, (4) a shift to the broader "defense transformation" and its partial empirical investigation in the early 2000s, and (5) second and third thoughts questioning the RMA paradigm from 2005 onwards. Since then, the conceptual narrative has evolved further into a sixth wave, characterized by diffusion of military innovation in small states and middle powers, as well as in the strategic competition between great powers, particularly China, Russia, and the United States. While adopting a linear interpretation of the RMA may inherently present empirical difficulties, it is useful to modify and contextualize Gray's initial categorization to compare and contrast prominent RMA themes, identify the intellectual thrust in the varying theoretical approaches, and place their relevance into a broader context of defense and security policy-making processes (see Figure 2.1).

First wave: Soviet MTR

The first-generation RMA theories have been initially conceptualized by the Soviet strategic thinkers in the early 1980s. Soviet strategic discourse at that time focused on the emergence of the so-called military-technical revolution (MTR). According to Dima Adamsky, the origins of the Soviet MTR reflected two interrelated theoretical variations: (1) the impact of the scientific progress on the scope, scale, and methods of future military operations; and (2) a response to Western doctrinal and technological innovations, particularly the adoption of the 1982 U.S. AirLand Battle (ALB) and NATO's Follow-on Forces Attack (FOFA). The former emphasized theoretical discussion of the long-term implications of military scientific-technological break-throughs such as laser, kinetic energy, electro-optics, remote control, automated decision support systems, telecommunications, and PGMs on the future of warfare. The latter constituted a continuous effort to rethink Soviet operational concepts amid increasingly sophisticated "deep-strike" capabilities fielded by the U.S. and NATO forces, which were perceived as significant enough to shift the "correlation of forces" between the Soviet Union and the West. In this context, Adamsky argues that "the merging of the two discussions produced a cogent military theory, which, while relying in part on Western technological achievements, was nonetheless highly original" (2008, 262).

Indeed, throughout the 1970s and 1980s, the Soviets became increasingly aware of the potential of emerging technologies in the U.S. arsenal that created new military capabilities as force multipliers, threatening to leverage traditional Soviet quantitative advantages vis-à-vis the U.S. and NATO forces in Europe. As Murray and Knox noted, "the appearance in the 1970s of striking new tech-nologies within the American armed forces … suggested to Soviet thinkers that a further technological revolution was taking place that had potentially decisive implications for the Soviet Union … from the Soviet perspective this was a par-ticularly frightening prospect" (2001, 3). The Soviets were also alarmed by the military lessons of wars between Israel and its Soviet-armed Arab neighbors, in which radar detection and precision firepower combined with new technologies and weapons produced high attrition rates on both sides (Shimko 2010). At the same time, they studied the operational implications of the new ALB and FOFA doctrines, which stressed "initiative, depth, agility and synchronization" by attacking deep in the rear through a combination of stand-off precision fire, inter-diction, and ground offensive operations (U.S. Department of Defense [DoD] 1982). From a Soviet perspective, this inherently threatened the potential for Soviet forces to rely on their traditional strategy of multiple echelons/combined arms formations pushing forward on the battlefield. Recognizing these develop-ments and changes in the "correlation of forces," the Soviets, under the command of Marshal Nikolai V. Ogarkov, Chief of the Soviet General Staff from 1977 to 1984, began to intellectualize the contours of the emerging MTR. Their strategic thought increasingly pointed to the fundamental discontinuity in warfare stem-ming from new technological innovations. In particular, they based their views on the following assumptions and observations (Krepinevich 1992, 7):

1   The rate of technological change is increasing, placing a greater premium on the ability of military organizations to adapt quickly to remain competitive on the battlefield.
2   The ability to move information rapidly while denying the enemy that capability is becoming ever more important, perhaps decisive. Space-based communications systems are seen as being extremely important in this process.

3 The "electronization" of warfare is proceeding apace, and modern warfare will witness the emergence of a new kind of operation: the "electronic-fire" operation.
4 Modern warfare is based on the delivery of extended-range nonnuclear strikes throughout an opponent's entire territory, destroying (or threatening to destroy) an opponent's key political, economic, and military targets.

Throughout the 1980s, Ogarkov voiced his views on future warfare in a number of professional periodicals as well as monographs, arguing that the MTR may render traditional Soviet operational art and strategy obsolete and stipulate a major discontinuity in military affairs in which quality is far more important than quantity. Ogarkov believed that advanced technologies – conventional PGMs coupled with enhanced sensors – would pave a way for qualitatively new and incomparably more destructive forms of warfare than ever before, diminishing the role of nuclear weapons in future wars. In his perspective, the effectiveness of new conventional weapons combined with "informatics" or advanced command, control, and reconnaissance systems would essentially correspond in magnitude to strategic or political effects as tactical nuclear weapons (Petersen 1984, 34). For example, in a 1985 book titled *Istoriya Uchit Bditelnosti* (*History Teaches Vigilance*), Ogarkov noted:

> A profound, and in the full sense of the word, revolutionary change in military affairs is continuing in our day in connection with the further development and qualitative improvement of nuclear weapons, rapid development of electronics, and in connection with the significant qualitative improvement of conventional means and methods of armed conflict.
>
> This, in turn has a decisive influence on all other aspects of military affairs, most of all on the development and improvement of forms and methods of military operations and, consequently, on the organizational structure of troop units and naval forces and improvement of control systems and organs.

Moreover, Ogarkov emphasized that the future battlefield would merge traditional conceptions of the front and rear areas with new weapons technologies and information systems allowing a near-simultaneous engagement of entire array of targets at greater distances, precision, lethality, and speed. The increasing value of space-based systems, unmanned systems, and automated detection and engagement integrated in a network of networks would dramatically redefine linear concepts of warfare. Amid these changes, the Red Army would have to rethink its operational concepts, adjust force structure, and redefine methods of waging war in each military service. As Adamsky noted, "Soviet theorists argued that given the tendency toward greater mobility and deception, the time available for destroying a target once it was identified would be limited. Thus, there was an acute need to develop an architecture that would consolidate the reconnaissance systems with high precision, fire-destruction elements, linked through the command and control channels" (2008, 257).

These efforts essentially shaped the development of two operational concepts: (1) Reconnaissance-Strike Complexes (RUK; in Russian, Рекогносцировочно-ударный комплекс) and (2) Operational Maneuver Groups (OMG). Both concepts were essentially doctrinal responses to the Western "deep-strike" ALB and FOFA doctrines, projecting a Soviet "deep-strike" version – an integrated mix of long-range fire systems, information systems, and command-and-control systems in a "network of networks" capable of engaging "a wide array of critical targets at extended ranges with a high degree of accuracy and lethality" (Krepinevich 1992, 6). In theory, the "RUK" would allow "simultaneous engagements of the enemy throughout the entire depth of his deployment ... capable of destroying small, mobile, targets with the use of long-range, high-precision munitions in combination with area sensors and automated command and control" (Watts 1995, 2). As the internal politico-economic conditions of the Soviet Union rapidly deteriorated in the late 1980s, leading to the eventual collapse of the Eastern Block and dissolution of the Soviet Union, the MTR remained solely theoretical and conceptually focused on technological over operational and organizational factors. However, while the Soviet Union lacked the technical capabilities and financial resources to pursue the MTR, its seminal military writings on the MTR at that time projected far more coherent understanding of its scope and implications than in the West – from abstract thinking to definitions of its sources, elements, and long-term consequences (Naveh 1997). Therefore, the credit for the intellectual discovery of the MTR and later the RMA is largely given to them (Marshall 1992).

Second wave: RMA and its early adaptation in the West

During the early 1980s, the U.S. intelligence community identified, monitored, and analyzed changes in the direction and character of Soviet strategic thought. The CIA at that time disseminated a number of internal memos and National Intelligence Estimates (NIEs) that pointed to new developments in Soviet weapons, changes in force structure, and new combined-arms approach in the Soviet doctrine, including Soviet RUK operational concepts – interpreted as an integrated triad of reconnaissance and target acquisition complexes, automated command-and-control elements, and long-range striking systems (CIA National Foreign Assessment Center 1981; CIA Office of Soviet Analysis 1984, 1987, CIA Directorate of Intelligence 1989). Indeed, the CIA was also able to clearly discern Soviet concerns of the U.S. ability to shift the military balance in the West's favor by exploiting superiority in weapons technologies. Generally, however, these assessments downplayed the overall conceptual significance and validity of the MTR, portraying it as an incremental or evolutionary modernization of individual systems, constrained by Soviet technological, manpower, and resource limitations. For example, in a 1985 NIE titled "Trends and Developments in Warsaw Pact Theater Forces," the CIA Office of Soviet Analysis projected that "while we see a new series of weapon systems now beginning to enter the force, these generally have modest improvement in firepower, mobility, and survivability ... Through the technology is new and significant for Soviet forces, it does not constitute a breakthrough as much as a 'catch-up' to technology available in the West ... In the near term, there is a little prospect for dramatic change." Others dismissed the MTR as purely "propagandistic hyperbole" (Blaker 1997, 5).

In the early 1990s, however, a small group of U.S. defense analysts at the Office of Net Assessment (ONA) in the Office of the U.S. Secretary of Defense began to systematically observe and evaluate Soviet MTR writings. Specifically, Andrew W. Marshall and his protégés, including Andrew Krepinevich, recognized the significance of the MTR discussion. In 1990, the ONA initiated a major assessment of the Soviet MTR writings to ascertain its existence and highlight the most important strategic management issues and operational and organizational policy imperatives. According to Marshall, many panels were formed to identify critical issues – that is, what was really new, produce a shared language and vocabulary for discussing the character of the changes in warfare, and build a policy-relevant consensus that the U.S. military was indeed in a period of major change. The resulting 1992 report, *The Military-Technical Revolution: A Preliminary Assessment*, confirmed the validity of Soviet MTR theories as a fundamental discontinuity in the conduct of military affairs. In the report, Marshall and Krepinevich argue that the significance of the MTR is not only in the technological aspects or the speed in which change takes place but rather in the *magnitude* of the change itself, which "at some point ... will invalidate former conceptual frameworks by bringing about a fundamental change in the nature of warfare and, thus, in our definitions and measurement of military effectiveness" (Krepinevich 1992, 3). In other words, Soviet understanding of the MTR has been too narrow – limited to the impact of new technologies and weapons systems on existing concepts of warfare, which mitigates the significance of other organizational, operational, and human drivers of change. While technological change is a necessary, it is not a sufficient factor to neither propel a disruptive military change nor guarantee military success. Thus, MTR is no longer adequate. With this assumption, Krepinevich and Marshall redefined the MTR into RMA as "the application of new technologies into military systems combined with innovative operational concepts and organizational adaptation that alters fundamentally the character and conduct of military operations" (Krepinevich 1992, 3). This new definition inherently transcended the first generation of Soviet MTR, suggesting the interplay of four key drivers: (1) technological change, (2) military systems development, (3) operational innovation, and (4) organizational innovation (Figure 2.2). "When (and only when) these elements are combined," they argued, "a dramatic improvement in military effectiveness and combat potential would take place" (Krepinevich 1992). Their study concluded with the view that the United States is at the *beginning stage* of a period of such revolutionary change in warfare.

From the early 1993, following the lessons learned in the Gulf War (1991), which demonstrated the potential of an entire catalogue of new sensors, platforms, and munitions, the larger contours of the RMA debate emerged – centering on its factual validity, precedence, and implications predominantly for the U.S. military. In particular, the RMA discourse during this period reflected the following four

Technological Change:	Military Systems Development:	Operational Innovation:	Organizational Innovation:
The increasing ability to gather, process, and disseminate information; major improvements in the range, accuracy, and lethality of conventional munitions; advanced simulations systems; and joint operations and network integration.	The integration of new technologies into military systems or munitions; use of space platforms, low-observable or stealth systems, electronic warfare systems, variety of sensors to provide near-real time targeting information.	Dramatically different operational concepts that would closely complement technological advantages. These would include joint and combined operations, information dominance, space/ air/sea/land control, surgical strikes, strategic mobility, unconventional warfare, etc.	Inter-service integration; highly-agile and flexible acquisition system; organizational peacetime innovation. These include the most difficult yet important parts of the transition. Large-scale organizations – especially military organizations are often highly resistant to change.

Figure 2.2 Elements of revolutions in military affairs

Source: Adapted from Krepinevich (1992).

themes: (1) the historical role of technology as a contributor of military effectiveness in war; (2) the sources and nature of military innovation; (3) the interpretations of processes of change brought by the information revolution; and (4) policy implications linked with short- and long-term consequences of the RMA for military organizations. In this context, the historical approaches to the RMA seemed to dominate the narrative, pertaining to the interplay of innovation, technology, military effectiveness, adaptation, and change in warfare in time (Murray 1997). Some of the pressing questions include: What is the (historical) evidence for RMA theory? When, how, and why do RMAs happen? How do RMAs "work"? What difference do RMAs make in the course of strategic history? During this phase, two distinct terms concomitant with major military change began to appear (sometimes interchangeably) in the RMA discourse: *MRs vs. RMAs*. MR is often accredited to Michael Robert's 1955 inaugural lecture titled "The Military Revolution, 1560–1660" at the Queens University Belfast. Roberts focused on the sixteenth-century European military history, seen as a time of major change in warfare and military organization. He proposed a single MR consisting of four distinct elements: (1) revolution in tactics, (2) revolution in strategy, (3) increased intensity in the scale of warfare, and (4) the impact of war on society (Parker 1976). His thesis sparked a debate among a small group of historians on whether there was a MR and, if so, when it occurred and what form it took. Notwithstanding the non-conclusive nature of this debate, the term MR began to signify a larger *sociopolitical and military* paradigm shift in how society, the state, and military organizations prepare and conduct war. According to Murray and Knox, "military revolutions recast society and the state as well as military organizations. They alter the capacity of states to create and project military power ... They [are] uncontrollable, unpredictable and unforeseeable" (2001, 7). In other words, MRs may reflect a grand strategic dimension of military change, whose effects extended beyond the battlefield and military organizations. On the other hand, within or alongside the cataclysmic MRs are lesser RMAs characterized by Murray and Knox as "periods of innovation in which armed forces develop novel concepts involving changes in doctrine, tactics, procedures, and technology ... RMAs [also] take place almost exclusively at the operational level of war. They rarely affect the strategic level, except in so far as operational success can determine the large strategic equation. RMAs always occur within the context of politics and strategy – and that context is everything" (2001, 12).

In many ways, the discussion on the MRs and RMAs has been shaped by Alvin and Heidi Toffler's *War and Anti-War* (1993), which viewed military innovation as a function of broader socioeconomic changes, as explained in their theory of *The Third Wave* (1980). The *Third Wave* argues that historical military-societal-economic-and-political transformations or power shifts result from new types of productive activity and new sources of wealth creation (Toffler 1984, 125–131). In this context, Alvin and Heidi Toffler conceptualized the evolution of modern warfare into three waves based on particular inventions in agriculture (First Wave), industry (Second Wave), and knowledge or information (Third Wave) that significantly changed the character of warfare. Their central premise is the notion that MRs are dependent on overall or cumulative paradigm shifts in the society, economy, politics, and technology:

> A military revolution, in the fullest sense, occurs only when a new civilization arises to challenge the old, when an entire society transforms itself, forcing its armed services to change at every level simultaneously – from technology and culture to organization, strategy, tactics, training, doctrine, and logistics. When this happens, the relationship of the military to the economy and society is transformed and the military balance of power on earth is shattered.
>
> (1993, 32)

Alvin and Heidi Toffler made an important contribution to the RMA literature by recognizing the importance of the broader socioeconomic context shaping the emerging information revolution – its socioeconomic, political, and strategic drivers in which technology serves as a necessary but not sufficient factor in propelling particular discontinuities in the modes of warfare. They viewed the shift from the industrial to information age "war forms" parallel with the evolution of particular "economy forms." This theme reverberated in many subsequent works and publications. For example, Krepinevich's (1994) seminal article in the *National Interest* titled "Cavalry to Computer: The Pattern of Military Revolutions" displayed the symbiotic relationship between MRs and RMAs (see Figure 2.3). In the article, Krepinevich identified at least ten MRs since the fourteenth century, which propelled the emergence of particular RMAs.

Military Revolution	Technological Change	Operational Innovation	RMA Change in Conflict
Infantry Revolution (1337–1453)	Six-foot yew longbow;	Archers integrated with dismounted men-at-arms;	Infantry displacing the dominant role of heavy cavalry on the battle field;
Artillery Revolution (1337–1453)	Lengthening of gun barrels (increase in range, rate of fire, and destructive force);	Artillery siege trains - capable of destroying fortified castles;	Gunpowder artillery displacing the dominance of the defense in siege warfare;
Sail & Shoot Revolution (1494–1650)	Ship design – floating artillery platform	Naval warfare – sail and shoot;	Change from oar-driven galleys to sailing ships with mounted large guns;
Fortress Revolution (16th cent.)	Defensive fortification with lower, thicker walls with bastions, javelins;	Trace Italienne;	Return of the static defenses;
Gunpowder Revolution (16th cent.)	Muskets;	New use of firepower Linear tactics;	Combination of artillery and musket fire;
Napoleonic Revolution (18th cent.)	Standardization of artillery calibers, carriages, interchangeable parts;	Field Armies - Mobility/Coordination Creation of the Division Staff System;	Large, organized armies performing mass frontal assaults preceded by long artillery preparations;
Land Warfare Revolution (19th cent.)	Railroads Telegraphs Repeating Rifles;	Strategic mobility and logistics;	Brief artillery preparation fires, infiltration tactics, use of light machine gun;
Naval Revolution (mid- 19th cent.)	Metal-hulled ships, turbine engines, long-range rifled artillery Steam propulsion, Submarines;	Submarine strategic blockade; Anti-submarine warfare;	Transformation of the character of war at sea;
Mechanization, Aviation, and Information (1918–1945)	Mechanization Radio/Radar Aircraft design Combustion engine	Blitzkrieg Carrier battle-groups Aerial bombardment	Modern amphibious/ mechanized armored warfare;
Nuclear Revolution (mid-20th cent.)	Mounting nuclear warheads on ballistic missiles (ICBMs)	Deterrence & Nuclear Doctrines (U.S. nuclear submarine force, Soviet Strategic Rocket Force)	Prospects of complete destruction of a state's economic and political fabric;

Figure 2.3 Patterns of military revolutions

Source: Adapted from Krepinevich (1994).

The theoretical debate on the origins, trajectory, scope, and character of MRs in the Western strategic thought led increasingly to normative calls to substantially conceptualize the RMA within policy-oriented context. The practical RMA debate in the early 1990s divided U.S. decision makers into two contending groups: (1) those who focused on the external perspective of the RMA as a means of attaining strategic objectives in the evolving geostrategic environments and emerging security challenges of the post-Cold War era, and (2) those viewing the RMA through internal processes as an organizing principle or a tool that can shape and determine future policy, acquisition programs, resource allocation, and bureaucratic relationships. In this context, Cooper argues that "the most important determinations that must be made concerning the RMA initiative are not analytical (epistemology), but of purpose (teleology)" (1994, 99). This means defining the strategic purpose and relevance of the RMA in a broad spectrum of conflict types, while determining its potential military benefits at the operational and tactical levels, strategic implications and consequences, and the question whether pursuing RMA is the most appropriate instrument for addressing evolving problems. In short, find ways how to best exploit and implement the RMA.

A number of studies focusing on this problem emerged at that time. For example, Michael Mazarr's 1994 book titled *RMA: Framework for Defense Planning* viewed the integration of new information and communication technologies in the military as both drivers as well as outcomes of a larger societal transformation, urging American policy makers to rethink traditional concepts of defense planning and management. Mazarr pointed to the relative absence or no sustained discussion of the nature and implications of the RMA in the official U.S. defense planning documents at that time, particularly the *Bottom-Up Review* (1993), which he argued reflected classical example of military leaders planning to fight the last war (Mazarr 1994, 6). Libicki and Hazlett in "The Revolution in Military Affairs" (1994) similarly argued that the most fundamental strategic challenge for U.S. defense planners is to convert the technological aspects of the MTR into an organizational and operational RMA. The core of the RMA problem, in their view, is whether the U.S. military should place a greater emphasis on managing short-term, known military tasks or tackle largely unknown tasks in the decades ahead (Libicki and Hazlett 1994).

In retrospect, the early adoption of Soviet MTR concepts and its subsequent adaptation into an RMA in Western strategic thought during 1990–1994 reflected a broader search for a new strategic paradigm that would account the prevailing transition into a post-Cold War era. On one hand, the output of RMA studies at that time was not comprehensive and as a result offered only limited or broad policy choices (Metz and Kievit 1995, 2). At the same time, however, the intellectual thrust to explore both MRs and RMAs – their origins and sources, scope and dimensions, dynamics and patterns, external implications, alternatives, and policy imperatives – became evident. A number of contending conceptual and argumentative differences emerged, ranging from transhistorical grand theories attempting to identify previous MRs across time and space toward theoretical frameworks trying to define the elements, structure, and impact of the contemporary RMA.

Accordingly, the RMA debate centering on the continuity and discontinuity in the means and methods of warfare, feasibility and utility of RMA, and subsequent normative calls to rethink policy and strategy gained rapid momentum. Its unifying theme was that the RMA transcends beyond the sole dimension of military technologies or systems and involves complex organizational and operational issues, which need to be resolved in order to effectively exploit the RMA and its potential.

Third wave: RMA "technophilia"

During the mid-1990s, the RMA became an "acronym of choice" in the U.S. defense planning and strategic studies community (Murray 1997, 64), "arousing tremendous excitement among American defense planners" (Metz and Kievit 1995, 1). The analytical thrust in the RMA literature shifted from identifying historical patterns of MRs or RMAs toward future-oriented defense policy imperatives on *how to increase military effectiveness at reduced cost*. RMA proponents generated a lexicon of "revolutionary" vocabulary and operational concepts that inherently reflected a growing sense to rethink most, if not all, underlying assumptions behind modern warfare. Key questions focused on how to redefine defense and technology management priorities, what changes are needed in the organizational force structures, acquisition plans and procurement programs, and, perhaps most importantly, how to link the RMA into a viable operational framework. In short, the fundamental theme driving the RMA narrative and debate at that time could be summarized thus: *what should be accomplished by the RMA, and how it should be implemented?*

The idea to develop radically different concepts of operations, organizations, doctrines, and practices based on the RMA increasingly shaped the research agenda in the U.S. DoD. For example, in January 1994, then Deputy Secretary of Defense, William Perry, established the "RMA Initiative" – a group of task forces within the U.S. DoD to coordinate RMA-related research and activities at inter-service levels. Its aim was two-fold: (1) to gather data, define most plausible strategic environment for the years 2010–2015, and identify most relevant technologies and operational concepts; (2) to conduct simulation exercises or war games to assess the impact of RMA in military operations and produce a report on the results (Galdi 1995). The RMA also began to stimulate research within individual military services. For example, Gen. Gordon Sullivan, then Army Chief of Staff, was widely perceived as one of the pioneers of the RMA in the U.S. Army (Sullivan and Coroalles 1995). In March 1994, Sullivan released a memo concerning the development of what he called *Force XXI: Digitizing the Battlefield*, a comprehensive effort to rethink and redesign the U.S. Army from its traditional, fixed, Cold War posture and doctrine toward an information-oriented, networked, and dynamic force (Sullivan 1994, 2–4). In particular, *Force XXI* created a conceptual foundation of a new doctrine, operational concept, organizational force structure, training, and capabilities all of which centered on the ideas of interconnectivity, digitization, and information-based weapons technologies. It also established the "Army Digitization Office" responsible for translating *Force XXI* concepts into overall mission planning and execution at all levels of warfare (Galdi 1995).

One of the most influential RMA architects at that time was William Owens. In 1996, Owens, then Vice Chairman of the Joint Chiefs of Staff, conceptualized the *System-of-Systems* at the core of the RMA (Owens 1996). The system-of-systems concept focused on two elements: information and jointness. It envisioned integrating existing (and in some ways overlapping) inter-service platforms and components – particularly, advanced C4I systems with ISR systems into a coherent, interoperable joint framework. Connecting diverse C4I architectures and ISR information systems would then enable unprecedented situational awareness capabilities – so-called "Dominant Battlespace Knowledge" (DBK) – envisioned by Owens across a large area of operations (200-by-200-mile boxes). In conjunction with PGMs and their platforms, the system-of-systems approach would exploit DBK that would accelerate decision-making processes and result in a "precision force in action and results" (Owens 1996, 37) and essentially enable the U.S. military to locate, track, and destroy enemy forces with virtual impunity. The U.S. forces would be able to identify, assess, and track any target in nearly real-time and subsequently attack it by applying "the right kind of force, at the right time, against the right targets" (Blaker 1997, 10). According to Owens, "the key was seeing power in functional interactions and synergy" (2002, 57). In his view, the RMA might give the U.S. military the capacity to shape directly the strategic environment by deterring potential aggressors, as well as respond to a full spectrum of threats by using more lethal, agile, stealth, and focused force. Therefore, the U.S. military should essentially accelerate the process toward the RMA by setting new resource allocation priorities and embracing broad technological, organizational, structural, and doctrinal changes. In short, the system-of-systems would constitute a qualitatively different military potential – not only a military superiority but rather dominance over any opponent and capabilities.

Indeed, prior to the arrival of Owens as the new vice chairman of the Joint Chiefs of Staff in 1994, individual U.S. services viewed the RMA not as a disruptive change in projecting military power but rather as a continuing drive for technical improvements in terms of their own individual systems and roles. In other words, the RMA was more about modernization and reform than revolution. Some may even argue that despite the growing RMA rhetoric in military journals and publications in the early 1990s, the U.S. DoD was not keen in pushing the RMA. As Blaker noted, "those in charge of the U.S. military were riding the crest of victory, first in the Cold War, then in Desert Storm. They were convinced … that the quality and effectiveness of the military they had built over the previous two decades was very high … [accordingly] the U.S. military was content to avoid rapid change" (1997, 7). Furthermore, Owens' critics pointed to the unrealistic nature of system-of-systems, not only for its overreliance on technologies but also on its assumptions that other countries or adversaries are not able to deny the United States its information dominance through deception or concealment (Cohen 1996). Hence, the RMA idea faced reluctant skepticism from those who preferred traditional military concepts of power as well as those who supported the development of individual service-level RMA concepts.

Owens' RMA vision, however, challenged these views in two important aspects: (1) it offered a broad road map toward integrated technological and operational compatibility between ground, air, and maritime service components; and (2) it formulated a new set of priorities in defense management and planning, aiming to accelerate RMA-oriented processes. Procurement priorities would now focus on particular advanced systems and platforms that allow greater connectivity and interoperability, specific R&D programs that would accelerate the RMA technological drive, and developing "joint" operational concepts. It was at this stage that the RMA became increasingly publicized and institutionalized as evident from official DoD publications such as the 1996 *Joint Vision 2010* (*JV2010*) and the 1997 *Quadrennial Strategy Review*. The *JV2010*, in particular, published by the Office of the Joint Chiefs of Staff in July 1996, was interpreted as a conceptual template on how to channel Owens' system-of-systems concept into a joint operational framework. It envisioned the RMA as a means to achieve new levels of effectiveness across a range of military operations, suggesting that the U.S. forces will be "able to accomplish the effects of mass – the necessary concentration of combat power at the decisive time and place – with less need to mass forces physically than in the past" (U.S. DoD 1996, 18). In other words, military operations would be based to create "massed effects." To do so, the *JV2010* introduced four operational concepts: (1) dominant maneuver, (2) precision engagement, (3) full dimensional protection, and (4) focused logistics. The synchronized application of these concepts would enable the U.S. forces to gain a "decisive" advantage in movement, mobility, accuracy, visibility, engagement, and control of a range of military operations – from peacekeeping to high-intensity combat. The desired end-state is then conceptualized as a "Full-Spectrum Dominance" capability (U.S. DoD 1996).

According to Mitchell, while the *JV2010* "represented the distillation of 20 years of technological advance and operationally focused thinking in the U.S. armed forces" (, 34) with its elaborate conceptualization of joint maneuver warfare and information superiority, its operational context was fundamentally related to integrating new technologies into existing platforms, organizations, and doctrines based on the 1980s' ALB. In other words, *JV2010* was more about a linear extension of previous concepts and practices and did not account for the complexity of the challenges brought by the global information revolution – from the widening globalization and changes in business operations to changes in the society, changes in the character of conflicts, and overall changing nature of the sources and nature of power in the information age. To address these issues, the RMA as a distinct paradigm shift in warfare required a radically different approach. The tipping point came in December 1997 with the publication of the National Defense Panel's report titled *Transforming Defense: National Security in the 21st Century* (Odeen *et al.* 1997). Its authors asserted that security challenges of the twenty-first century will be "quantitatively and qualitatively different from those of the Cold War, and will require a fundamental change to U.S. national security institutions, military strategy and defense posture" (1). Therefore, the United States must embrace a broad transformation of its military and national security structures, operational concepts, defense management processes, and equipment. "The United States," the panel emphasized, "needs to launch a transformation strategy now" and ensure that DoD and the services "accord the highest priority to executing it" (4). The report also dismissed previous efforts to "balance" or modernize existing U.S. military capabilities, which are essentially not relevant to the long-term needs of the military (43). Accordingly, the techno-oriented RMA gradually morphed into a much broader policy-oriented process of defense transformation that aspired beyond changes in operational concepts, force structures, and equipment. Subsequent U.S. national security and defense planning documents, studies, and public statements by senior military and civilian leaders supported the concept (Metz 2006).

Shift to "defense transformation"

The Bush administration endorsed ideas and concepts associated with the RMA in a broader policy of "defense transformation" (Adams 2008). Speaking at the Norfolk Navy Base in February 2001, President Bush emphasized the need to "move beyond marginal improvements to harness new technologies that will support a new strategy … [and] begin creating the military of the next century" (Sanger 2001, 1). In the speech, as well as during his campaign trail, he called for lighter, more agile, and more lethal ground forces as well as both manned and unmanned air forces capable of striking across the globe with precision. The Bush administration realized that while emerging RMA technologies may enable defense transformation processes, new threats to U.S. security make transformation necessary (O'Rourke 2007). The rationale for pursuing the RMA thus shifted from technological revolutions by default (dominant in the early 1990s) to the changing strategic and operational imperatives – the emergence of complex, hybrid, and nonlinear threats involving terrorism, irregular warfare, insurgencies, and asymmetric warfare challenges. Soon after taking office, then the Secretary of Defense, Donald Rumsfeld, appointed Andrew Marshall – essentially the godfather of the RMA – to lead the team conducting the congressionally mandated 2001 *Quadrennial Defense Review* (*QDR*). The 2001 *QDR* presented the need to transform the U.S. military, stating that "a fundamental challenge confronting DoD is ensuring the U.S. forces have the capabilities they need to carry out the new defense strategy and meet the demands of the 21st century. Toward that end, it is imperative that the U.S. invests and transforms its forces and capabilities" (U.S. DoD 2001, 40). In other words, its main argument amplified the idea that the challenges and opportunities confronting the U.S. military are significantly different from those encountered during the Cold War or the 1990s' era and that the character of the threats is changing. Therefore, U.S. military capabilities must be inherently linked to these new challenges (Krepinevich 2002). The 2001 *QDR* also sharpened the definition and character of defense transformation (U.S. DoD 2001, 40):

> Transformation results from the exploitation of new approaches to operational concepts and capabilities, the use of old and new technologies and new forms of organization that more effectively anticipate new or still emerging

strategic and operational challenges and opportunities and that render previous methods of conducting war obsolete or subordinate.

Transformation can involve fundamental change in the form of military operations, as well as potential change in their scale. It can encompass the displacement of one form of war with another, such as fundamental change in the ways war is waged in the air, on land and at sea. It can also involve the emergence of new kinds of war, such as armed conflict in new dimensions of the battle-space.

After the 9/11 terrorist attacks, the concept of defense transformation has both accelerated and broadened into an all-encompassing umbrella for multifaceted initiatives. As Dombrowski and Ross noted, defense transformation became an "enterprise" – "a complex, risky, political-economic undertaking in which the U.S. government sought to enhance its military capabilities and the private defense-industrial sector sought to reap renewed profits from defense sales at home and abroad" (2008, 13). The process involved restructuring and rebalancing the U.S. forces to deal with the new threat environment, while revamping defense management processes and changing strategic culture. According to Bitzinger, "the transformation of the U.S. armed forces was initially promoted as nothing less than a fundamental shift in the way wars would be fought in the future. Nothing was sacred: every piece of defense dogma was on the table for debate and discussion – force structure, organization, equipment, budgets, doctrine, and strategy" (2006, 6).

On October 21, 2001, Donald Rumsfeld set up the Office of Force Transformation (OFT), headed by Vice Admiral Arthur K. Cebrowski. With its research, the OFT generated one of the main intellectual thrust behind transformative imperatives that subsequently shaped the character and direction of joint and service transformation plans and programs. Its staff included RMA "pioneers" such as Andrew Krepinevich of the Center for Strategic and Budgetary Assessments, who conceptualized "transformation" as a distinct term, arguing that it enables or leverages the RMA. Krepinevich often emphasized that defense transformation is *not* synonymous with RMAs, nor it represents a war-fighting strategy; rather it stipulates a vision of future warfare and defense management surrounding "a process that a defense establishment undertakes if it believes a military revolution is under way, or is potentially under way" (Krepinevich 2002). In this perspective, Krepinevich argued that defense transformation requires first and foremost a strategy for large-scale innovation, which would include new operational concepts, doctrine, capabilities, force structures, training, etc. Together with Arthur Cebrowski, he often pointed that transformation is a *continuing process* (rather than an end-point) of anticipating the future, creating new competitive areas, identifying and leveraging new sources of power (2003, 8).

Searching for new transformation strategies that would transcend the RMA, the DoD published a number of transformation-related documents and roadmaps. The 2003 *Transformation Planning Guidance* framed defense transformation as a "process that shapes the changing nature of military competition and cooperation through new combinations of concepts, capabilities, people and organizations that exploit our nation's advantages and protect against our asymmetric vulnerabilities to sustain our strategic position, which helps underpin peace and stability in the world" (U.S. DoD 2003, 3). Other notable transformation policy documents include *Elements of Defense Transformation* (2004), *National Defense Strategy* (2005), and the *QDR* (2006). During this period all services and agencies within the U.S. military have also developed the rationale and roadmaps supporting the overall transformation vision. The U.S. Air Force, U.S. Navy, and U.S. Marine Corp have conceptualized their transformation visions, explaining how they would leverage RMA-type capabilities and effectively project force in future conflicts and crises. For example, *Navy Sea Power 21* strategy (2004) defined the U.S. Navy shifting its operations toward more flexible and expeditionary deployments in littoral (i.e., near shore) waters, using highly capable, multi-mission destroyers, advanced cruisers, new types of mission ships with smaller crews (U.S. Department of the Navy 2004). Similarly, with the *Air Force Transformation Flight Plan* (2003), the U.S. Air Force would become more expeditionary, exploiting new operational concepts and technologies in order to significantly enhance its capabilities and transform into a "global reconnaissance and strike force" dominating air, space, and cyberspace (U.S. Air Force 2003). Meanwhile, the U.S. Army would be reorganized into modular, brigade-sized forces called Units of Action and deployed in more distant areas and adaptable to the needs of specific contingencies.

Taken together, defense transformation propelled a number of unprecedented force modernization and procurement programs, strongly supported by the U.S. defense industries. As Shimko noted, "there was a mad rush to portray everything as transformational. Systems in development for years were suddenly repackaged and repositioned at the vanguard of military transformation" (2010, 134). In other words, transformation programs emphasized a new generation (rather than improvements) of technological, doctrinal, and organizational aspects as well as their multidimensional, synergistic relationships between them (Dombrowski and Ross 2008, 13–38). The main characteristics of "transformed" force included mobility, agility, lethality, stealth, survivability, extended range, precision firepower, and networked synchronization (O'Rourke 2007). For example, in 1999, the U.S. Army initiated the "Future Combat System" (FCS) program, which later became integrated into the broader "Future Force" program (and subsequently cancelled). According to Adams, the development and timeline of the FCS has been the most ambitious modernization vision of the U.S. Army since World War II and the most expensive weapons program in its history – extending to the year 2030 and involving around 550 defense contractors and subcontractors, with total costs projected at US$ 300 billion (Adams 2008). The Future Force would replace the so-called "legacy systems" (such as the 70-ton M1 tank or the 35-ton Bradley fighting vehicle) with a networked formation of initially 18 (later 14) manned and later only unmanned platforms and subsystems that would be linked to various weapons, drones, robots, sensors and hybrid-electric combat vehicles, and soldiers. With the FCS, the U.S. Army would be organized as a "Modular Army" consisting of "Brigade Combat Teams," a lighter, faster, and high-tech ground force using a fleet of 15–18-ton manned vehicles that in theory could perform all the functions of the M1 tank, the Bradley, and conventional artillery with greater lethality and no loss of survivability. The FCS also envisioned rapid deployment anywhere in the world, with the ability to drop a fully supplied and ready-to-fight brigade in 36 hours, followed by the full division within five days and five more divisions deployed within 30 days (Adams 2008).

Central to the implementation of the various defense transformation plans at that time were two interrelated operational concepts and their variations: (1) *Network-Centric Warfare* (NCW) and (2) *Effects-Based Operations* (EBOs). Both concepts have essentially reflected relevant doctrinal guidelines on the future operational conduct and organizational force structure. Both concepts attempted to provide answers on how to effectively fuse information superiority concepts with precision firepower into networked modes of warfare at both strategic and operational levels. As with other aspects of the RMA debate, they have been subject of contending viewpoints and modifications with regard to the validity, reliability, and applicability. Notwithstanding the ebbs and flows of these debates, a brief overview of both concepts must be included in any etiology of RMA narrative.

Network-centric warfare

The origins of NCW date to the mid-1990s, when the Joint Staff at the U.S. DoD together with outside defense contractors and consulting firms (Booz Allen & Hamilton) began to conceptualize baseline assumptions of NCW. The early NCW concept was closely linked to Owen's "system-of-systems" idea of connecting sensor networks, command control networks, and shooter platforms that would simultaneously share information in a common operating environment. Publically, however, the concept has emerged in a seminal article by Vice Admiral Arthur Cebrowski and John Garstka titled "Network-Centric Warfare: Its Origin and Future" (1998) in the U.S. Naval Institute's magazine *Proceedings* (Cebrowski and Garstka 1998). In the essay, the authors argued that the ongoing information revolution has transformed leading transnational corporations, which gained significant competitive advantage by integrating innovative network-centric strategies into their business models and operations. Reflecting earlier notions by Alvin and Heidi Toffler, who argued that MRs result from broader socioeconomic transformations, Cebrowski and Garstka argued that by exploiting information technologies and leveraging information through networks, these companies increased the value of shared information while improving efficiency, collaboration, and making timely, adaptive, and accurate decisions in line with rapidly changing conditions in their business domains. Similarly to network-centric processes in the corporate and societal domains, Cebrowski and Garstka inferred that the information revolution is having a tremendous impact on the military domain, changing not only the tools of war but also the rule sets and organizational architectures under which military organizations operate. They described the process of military response to the information age as a broader paradigm shift from "platform-centric" to "network-centric" style warfare and compared the transformation impact to the French concept of the *levee en masse* during the Napoleonic era. The NCW argument is that increased connectivity or networking of geographically dispersed (yet highly agile) military assets and units would yield a high-level situational awareness, which would then accelerate and sustain the speed of command and enable the synchronization of complex activities at all levels of warfare. In other words, the networking behavior generated by a "network-centric" force could potentially translate into timely and more accurate responsiveness, which would mitigate risks associated with the "fog of war," reduce costs of military operations, and potentially result in exponentially increased combat power.

Cebrowski and Garstka's seminal essay set the stage for further exploration of the NCW idea and propelled a new direction in the intellectual thrust behind the American RMA. Its development was marked by a series of influential semiofficial monographs published by the DoD's Command and Control Research Program (CCRP). The first book in the series, written by Alberts, Garstka, and Stein and titled *Network-Centric Warfare: Developing and Leveraging Information Superiority* (2000), sharpened some of the main assumptions characterizing a network-centric force. In particular, the authors argued that NCW does not represent "the answer" to all adversaries nor that it will automatically lead to increased combat power. Rather, they pointed that the shared situational awareness generated through effective networking processes could maximize the "force responsiveness" and, in doing so, maximize combat power to shape the battle to one's advantage. In addition to the vastly improved performance of weapons technologies and sensors that provide novel capabilities, the key enabler would utilize networked and adaptive (rather than hierarchical and static) organizational architectures and their ability to leverage information superiority as a source of power (Mitchell 2006).

Attempting to refine the early NCW concepts, Alberts, Garstka, Hayes, and Signori published a second book in the CCRP series titled *Understanding Information Age Warfare* (2001), in which they presented a more developed theory of operations in networked environments. They begin by describing how information superiority is created – defining the relationship between information, knowledge, and awareness – and subsequently attempt to explain how particular NCW-oriented processes may enable to translate and consolidate information superiority into increased combat power. The authors propose a structure of information environment consisting of three interrelated domains: (1) *physical domain* – the physical location or environment of land, sea, air, and space in which combat and maneuver operations take place, where systems are attacked and defended and where action is directly observable and measured through both direct and indirect sensing; (2) *information domain* – the virtual environment, providing space for the creation, manipulation, flow, and sharing of information among different actors and technologies, as well as serving principally as a medium for communication, and is critical for command and control of military forces with regard to information quality, reach, and interaction; and (3) *cognitive domain* – the individual and collective consciousness, perceptions, attitudes, and beliefs of the actors participating in the network, where decisions are formed, evaluated, and taken based on the understanding and interpretation of the data from both the physical and information domains. In *Power to the Edge* (2003), Garstka and Hayes added a fourth *social domain* that transmits perceptions, assessments, and decisions developed in the cognitive domain. Together, the theory of NCW then proposes that "a robustly networked force improves information sharing and collaboration, which enhances the quality of information and shared situational awareness. This enables further collaboration and self-synchronization and improves sustainability and speed of command, which ultimately result in dramatically increased mission effectiveness" (OFT 2005, 7).

At that time, the conceptualization of NCW has been relentlessly promoted by Vice Admiral Arthur Cebrowski and his followers at the OFT in their publications and presentations. However, as with any other RMA-oriented concepts, the idea has been contentious since its inception. Probing questions emerged as early as 1998 and continue to drive the NCW/RMA debate, particularly on how to measure NCW "capability gains" – how such gains arise, to what extent they be maximized – while minimizing the opponent's capabilities. Thomas Barnett (1999), for example, questioned the effectiveness of NCW vis-à-vis near-peer competitors, its adaptability to military operations other than war, the potential for information overload, and too much speed that may lead to wrong decisions. Other NCW critics pointed to the risks of the overreliance on technology in strategic decision making and the continuing importance of human factor. Admiral William J. Holland argued that "the fog of war does not go away – it will appear in new and different forms" (2003, 73). Similarly, Edmund Blash warned of "technically codified concept based on the unproved premise that machine intelligence and analysis are superior and can be substituted for the soldier in the loop … information and networking alone are not substitutes for combat maneuver and massing" (2003, 56). Ultimately, critics pointed out that the requirements to fully implement NCW and achieve the combined effect of self-synchronization are high. In the physical domain, all elements of a military force must be connected together to achieve secure and seamless connectivity and interoperability; in the information domain, people and platforms must be able to access, share, and protect information to maintain information dominance over an adversary; and at the cognitive domain, forces must be able to use this shared information to develop and share awareness of their environment with other network participants (Alberts *et al.* 2001).

Effects-based operations

While the NCW provided the key operational concept guiding the development of capabilities for a "transformed force," the EBO attempted to provide systemic methodologies to shape the adversaries' behavior, perceptions, and ability to operate in a coordinated manner (Smith 2003). Proposed as an alternative, substitute, or expansion to the traditional attrition-style high-intensity warfare, EBOs would target key "cognitive" elements of enemy's systems and leadership through a coordinated set of actions – a full spectrum of economic, psychological, diplomatic, and combat activities. The objective would be to create a full range of cascading effects (direct, indirect, complex, or cumulative effects) that would ripple through enemy's systems and lead to a rapid collapse of its willingness and ability to fight, without the need to destroy the bulk of its military forces through a lengthy process of attrition. In this context, EBOs would include (1) collapse the will and cohesion of the enemy; (2) defeat the enemy's strategy rather than his armies; and (3) convincing the enemy's leadership to accept one's terms (Davis 2001, 12). While similar ideas have been a part of strategic studies in the past and utilized by militaries throughout history, EBO advocates suggest that in combination with accelerating technological capabilities, information management, and precision applications of force, EBOs would orchestrate new synergies that could result in greater efficiency and effectiveness of such operations. In other words, EBOs would aim to achieve mass of effects without the need for mass of forces, resulting in victory at lower cost of lives and resources (Smith 2006).

Much like the NCW, however, EBO-related theoretical frameworks and approaches reflect diverse views, schools of thought, and definitions. Its origin is credited to Col. John Warden, the architect of the Gulf War air campaign in which he devised an adaptive systems approach to planning and execution of air power that paralyzed Iraqi forces and undermined their will to fight. Warden's "Strategic Paralysis" model envisioned that command of the air could be achieved more effectively through *parallel* (simultaneous) attacks on the entire array of high-value objectives rather than sequential attacks (Fadok 1995). In this context, he developed a model of targeting that identified five distinct rings or centers of gravity – starting at the center ring, he identified leadership targets, then means of production, infrastructure, population, and fielded forces at the outer ring. Warden argued that targeting the centers of gravity in any of the rings *simultaneously* is more effective than traditional sequential attacking (one at a time) aimed to "roll back" enemy's defenses from outer to inner direction (Mets 1999). This implied that the number of sorties flown would be relatively less relevant compared to their planning, coordination, and ability to achieve specific effects, which were not bound to absolute destruction of target lists (Deptula 2001a,c).

In 2001, U.S. Air Force theorist David Deptula expanded Warden's predominantly Air Force-oriented EBO model of parallel warfare into a broader defense transformation concept. Specifically, he argued that "parallel warfare is a manifestation of the revolution in military affairs, and effects-based operations are a critical enabler. More than a methodology for applying new technology, effects-based operations call for a basic realignment in war planning" (Deptula 2001b, 17). In this context, Deptula envisioned a fundamentally new approach to the application of force – linking the EBOs to every medium and element of warfare guided by national strategy. Specifically, he proposed an alternative operational concept of war based on control of the enemy rather than traditional methods of physical destruction, annihilation, and attrition. Controlling – preventing the intended use of adversary's critical systems that would inhibit his ability to operate – would be relatively more effective than the physical destruction or attrition of his forces. In other words, destruction should aim at achieving effects on enemy systems and not necessarily at destroying these systems. He concluded that "effects-based operations have the potential to reduce the force requirements, casualties, duration of conflict, forward basing, and deployment of forces previously required to prevail in war. In short, the parallel approach changes the basic character of war" (Deptula 2001b, 25).

Deptula's work prompted other U.S. defense research institutions to initiate further concept development and experimentation. In the 2001 white paper on EBOs, for example, the U.S. Joint Forces Command (J9) attempted to form a baseline consensus to EBOs by defining it broadly as "a process for obtaining a desired strategic outcome or effect on the enemy through the synergistic and cumulative application of the full range of military and non-military capabilities at all levels of conflict" (5). An "effect" would be synonymous with the physical, functional, or psychological outcomes, events, or consequences that result from specific military or non-military actions. EBOs would be then viewed as adaptive processes, beginning with developing knowledge of the adversary, the environment, and all available capabilities – that is, diplomatic, information, military, and economic. The knowledge would be used to determine the required "effects" needed to convince or compel the enemy to change behavior through coordinated and synergistic operation at strategic, operational, and tactical levels. In 2001, Paul Davis added a "probability dimension" to the EBOs, defining the concept as "operations conceived and planned in a systems framework that considers the full range of direct, indirect, and cascading effects, which may – with different degrees of probability – be achieved by the application of military, diplomatic, psychological, and economic instruments" (7). Subsequently, a large number of studies emerged focusing on multiple themes related to EBOs – from strategic to operational levels of warfare, that is, effects-based joint operations, effects-based planning methodology, effects-based net assessments, modeling, targeting, and others. However, after nearly eight years in conceptual development and a number of attempts to codify a working template or unified definition, EBO approaches remained unproven and fragmented in multiple domains and debates. "The 'concept' of EBO has remained largely just that – a conceptual construct" (Vego 2006, 51). Its critics pointed to the lack of clarity of the concept, ambitious goals, complexity, unreliability, unpredictability, and historical precedence that challenged the validity of the concept. The mounting criticism centered on the idea that EBO approaches are highly complex and impossible to measure and achieve in a continuously changing security environment. The resulting "EBO debate" reflected two critical questions: (1) How, if at all, do EBOs fit into the existing military lexicon and doctrine? (2) How should EBOs be planned and executed amid their inherent dynamic complexity that prevents clear answers to traditional tests of ways, ends, and means?

In particular, EBO critics argued that all operating environments are dynamic with an infinite number of variables, which essentially precludes an accurate scientific prediction of the outcome of an action. Therefore, EBOs cannot clearly anticipate reactions of complex systems (i.e., leadership societies, political systems), which are highly adaptive and have different behavioral mindsets. Its critics also emphasized that EBOs assume an unattainable level of knowledge of the enemy. Accordingly, at the operational level, EBOs may fail to provide clear and timely direction to staff, using terminology that is difficult to understand. Amid these growing concerns, the Joint Forces Command (JFCOM) headed by Marine Gen. James Mattis issued a memorandum in 2008 stating that "effective immediately, USJFCOM will no longer use, sponsor, or export the terms and concepts related to EBO... in our training, doctrine development and support of Joint Professional Military Education." The memo noted that "we must recognize that the term effects-based is fundamentally flawed, has far too many interpretations and is at odds with the very nature of war to the point it expands confusion and inflates a sense of predictability far beyond that which it can be expected to deliver" (106).

Second and third thoughts: modernization-plus

Notwithstanding the strong commitment and efforts by the Bush administration to implement the various defense transformation initiatives, as reflected in the 2006 *QDR*, the perceptions and viewpoints toward the RMA began to shift from the mid- to late 2000s. According to Richard Bitzinger, "the challenge for the U.S. military has been translating this transformational vision into a credible and effective set of capabilities, strategies and organizations" (2006, 12). Indeed, since its inception, defense transformation has arguably reflected an extremely ambitious and diversified agenda for change for the U.S. military. Transformation advocates visualized a broad spectrum of new technologies and unprecedented capabilities. Unmanned systems would permeate and operate in multiple battlespace domains (air, land, sea, and space), and ubiquitous sensors would link with the next generation of weapons systems and platforms. The navy would use smaller, faster, lighter ships with new stealthy hull designs; the air force would use unmanned drones and hypersonic crafts flying between the atmosphere and space. Future weapons would utilize new forms of kinetic energy, nanotechnology, biotechnology, and non-lethal technologies. Future operations would be network-centric rather than platform-centric, with networked forces yielding distributed effects and precise fires rather than mass. Cyber warfare and information operations would rival conventional operations (Dombrowski and Ross 2008). However, in the late 2000s, such ambitious rhetoric have culminated into a ubiquitous "silver-bullet" term that its all-purpose idea has outpaced its actual implementation.

Even before that, critics pointed out that the RMA thesis is untested – evolving "from exposition to consideration for implementation as a U.S. government policy so quickly that it outpaced the ability of scholarship to examine its underlying premises and evidence" (Biddle 1998). The various RMA-related defense transformation visions created high expectations that far exceeded available capabilities, resources, and realities. Transformation critics pointed to its unfulfilled promises and ambiguities as an open-ended process, suggesting the concept has turned into an empty slogan, cliché, or a buzz-phrase. The rationale for "new way of thinking and a new way of fighting" justifying virtually every defense initiative or proposal, whether RMA-related or not, signaled disorientation rather than a clear strategy (Freedman 2006). RMA skeptics further warned about single-dimensional or "locked-in" solutions to complex strategic challenges, overreliance on technological means that may be offset by adaptive enemies, and unacceptable risks in ignoring current or near-term defense requirements. Others warned that defense transformation is a misguided or even a dangerous concept as it is driven by the military's weapon system preferences and capability sets rather than policy imperatives (Raynolds 2006).

In large part, defense transformation and its proponents have also been undermined by the challenges and operational demands in the conflicts in Afghanistan and Iraq, which turned into protracted and complex counterinsurgency campaigns. Indeed, the priorities, resources, and focus in the American military establishment have inherently shifted from pursuing relatively open-ended transformational visions to tackling insurgent strongholds, cascade of multifaceted attacks, sectarian violence, and immediate tactical threats such as IEDs (Adams 2008). Inherently, the U.S. military became confronted by a spectrum of "hybrid" political, socioeconomic challenges of a non-linear conflict that it was not prepared for, nor had it anticipated. As Keith Shimko noted, "fixated on technology, [the U.S. military] was unable to conceive of wars and missions that were not primarily about guiding munitions to targets with incredible precision from great distances, and the military was ill-equipped and poorly trained for anything other than high-intensity inter-state warfare" (2010, 203). The confluence of various operational requirements in the counterinsurgency warfare in both Iraq and Afghanistan demonstrated the limitations of RMA technologies, particularly in winning the "hearts and minds" of the people. Amid these developments, transformation strategies and concepts began to reinforce ideas related to the "Second-Generation Transformation" (Farrell, Terriff, and Osinga 2010).

In 2006, as Robert Gates succeeded Donald Rumsfeld as Secretary of Defense, the overall U.S. defense transformation plans, leadership, management, experiments, exercises, and innovation drive changed its focal plane from the previous RMA-oriented discontinuous and a disruptive paradigm shift to a "shift in emphasis." In October 2006, Gates dissolved the OFT and transferred its research projects to other agencies within the department. Meanwhile, the DoD selected the JFCOM, a unified military command with a staff of more than 800 headquartered in Norfolk, as the military's premier "transformation laboratory" (O'Rourke 2007). More importantly, the DoD became increasingly concerned with the rising acquisition costs of key "transformation" weapons and systems coupled with their defense management problems. For example, between April 2003 and September 2004 alone, the total program costs for the U.S. Army's FCS escalated 35.2 percent, and research and development costs increased by 50.8 percent (Cordesman and Frederiksen 2006). During the same period, the acquisition timeline increased by 52.7 percent to 139 months. By 2005, the Army estimated that the total cost of FCS could easily reach $145 billion, some $53 billion more than originally estimated. Cordesman and Frederiksen noted that "the very real advantages of technology and modernization threaten to be lost by a massive procurement program that is a cost escalation nightmare, that forces constant cuts in both the active force and the numbers of new systems to be procured, and which can only be financed – even on paper – by cutting the funds for technology in the out years as well as limiting other key expenditures like operations and maintenance costs and manpower" (2006, 8).

As a result, from 2005 onward, the transformation debate shifted to issues of feasibility, affordability, and, ultimately, desirability of "transformational" weapons, platforms, and systems. The RMA discourse became less "revolutionary" or "transformational" in all its three key aspects – magnitude, speed, and implications. More critical voices emerged questioning both its projected efficiency and effectiveness. Dombrowski and Ross contend that the institutionalization and actual substance behind defense transformation visions, plans, and programs within the U.S. military establishment has reflected more of an *evolutionary change* or "modernization-plus" rather than a discontinuous, disruptive innovation:

> The U.S. transformation enterprise thus far falls short … while the visions promise discontinuity and disruption, plans and programs support only incremental, sustaining advances. Technological generation-skipping is nowhere to be found. Doctrine development is more linear than nonlinear. Organizational change features evolution and adaptation rather than recreation or, even, restructuring. Unless the gap between visions and plans and programs can be bridged, transformation is fated to be little more than routine modernization.
>
> (2008, 23)

Notwithstanding the inherent changes in the transformation rhetoric and its course in the U.S. armed forces, American RMA concepts, technologies, and lessons learned have shaped doctrinal development and military modernization processes in other select militaries around the world. Studies on RMA diffusion, emulation, and adaptation outside the United States that began to emerge in the late 2000s explored the extent and trajectory of military modernizations of select cases drawn from the European (NATO) states, China, Russia, Australia, and analyzed their relevance for the United States. Simultaneously, there has been a growing interest in exploring the RMA diffusion through the lens of global defense industries (Bitzinger 2008). Hence, the RMA debate arguably continued to evolve and resonate in theoretical and policy-oriented studies on military innovation, national security policies, defense economics, and future of warfare.

Toward a sixth RMA wave?

Based on the chronological overview of the RMA, it is clear that the RMA reflects a variety of perspectives, institutional impressions, strategic imperatives, operational concepts, organizational interests, and preferences. Arguably, one could contend that the rapid pace and sophistication of technological advances in the military domains coupled with changes in the character of conflict in the twenty-first century further shaped its conceptual diffusion, adaptation, and integration, notably in the United States. Accordingly, there has been no single RMA school of thought but rather a range of contending views and debates that attempt to explain the change and continuity of warfare, sources of military innovation, and resulting defense management and security policy imperatives. In a broader perspective, the resulting RMA debate has centered on five key aspects: (1) definition, metrics, and impact of the revolution or change in itself; (2) the pace, direction, and cost of technology in warfare, (3) the normative extent of the defense policy-making processes and resource allocation imperatives to implement the changes; (4) the actual and projected effectiveness of the RMA in the use of force; and (5) the sources, character, and pace of RMA diffusion and adaptation. In particular, the primary questions driving the RMA debate through its various stages include: Is there an RMA as a discontinuous paradigm shift in warfare and, if so, what does it mean? What aspects of warfare are being revolutionized? How are the character and conduct of war changing? Is the pursuit of RMA necessary and, if so, why? How do we define RMA capabilities – what military systems and platforms can be defined as RMA-oriented or transformational? Is the RMA sufficient to counter the progressive complexity of asymmetric security threats? Is the RMA affordable? How can modern military organizations implement and exploit the RMA? What are the key defense management challenges and policy limitations in pursuing the RMA? What are the key externalities and trade-offs in pursuing the RMA? How will the RMA shape traditional defense industries? What is the future of military innovation and modernization? These questions have inherently created a range of views, images, and arguments filtered through different theoretical lenses, focus areas, and methodological approaches. Moreover, they show the progression of the RMA debate from the existence and validity of the RMA, its meaning, levels, causes and effects, to its diffusion and adaption processes.

Notwithstanding the ebbs and flows of RMA debate and key contributions that have shaped theoretical discourse in strategic studies as well as defense policy-making processes over the last two decades, its significance continues to be unabated. The varying approaches to RMAs inherently converge (to a varying degree) into a continuous, conceptual, technological, organizational, and operational innovation aimed to dramatically amplify military capabilities and their effectiveness. As Murray noted, "RMAs involve putting together the complex pieces of tactical, societal, political, organizational, or even technological changes in new conceptual approaches to war" (1997, 63). When embedded in broader MRs, RMAs may lead to profound changes in the framework of war, altering the states' capacity to create and project military power for political ends. This of course depends on a number of variables, including the state's ability to recognize, anticipate, exploit, and sustain a comprehensive RMA-oriented military innovation. Therefore, understanding the RMA dynamics – their context, character, diffusion, adaptation, and impact – is fundamental to understanding modern security questions and defense management problems. As this chapter argued, the intellectual development of the RMA and its institutionalization into military organizations, predominantly in the United States, has evolved over time through at least five distinct phases. Notwithstanding its conceptual roots in the Soviet military thought and strategy, the RMA debate has been extensively an American debate, driven by U.S. military-technological innovation and operational experience. Yet, systemic research on the diffusion, adaptation, and impact of RMA-related concepts and technologies on strategic thinking, defense management processes, and security debates in other geostrategic settings has been relatively limited, emerging only in recent years. In other words, the key question is whether the continuing RMA debate may spin-off into a "sixth wave" – that is, on how have small states and middle powers – whether U.S. allies, strategic partners, or potential adversaries – responded or are responding to the RMA over the past two decades? In a reverse mode, to what degree have select small states actually shaped the RMA diffusion through their military-technological innovation, operational experience, and lessons learned? These questions may propel the "sixth RMA wave" toward which this book inherently aspires.

At the same time, however, one may also argue that the sixth RMA military innovation wave will continue its track in the military rivalries and strategic competition between great powers – particularly China, Russia, and the United States. Indeed, China's current political and military elites under President Xi Jingping believe that a new wave of the global RMAs is gathering pace, led principally by the United States, and China must therefore accelerate the pace of its military development. China's increasing power projection capabilities embedded in the People's Liberation Army's growing technological developments are gradually redefining regional military balance and subsequently U.S. strategy. In particular, China's asymmetric "counter-intervention" concepts and weapons technologies – designed to deny U.S. forces and its allies the freedom of action in China's "near seas" by restricting deployments of U.S. forces into theater (anti-access) and denying the freedom of movement of U.S. forces already there (area denial) – amplify the magnitude of strategic and operational challenges for U.S. forces in the region. In this context, numerous books and studies have been published in recent years on China's military modernization concomitant with the development of Chinese defense science and technology and the search for innovation in China's defense economy. This literature attempts to identify, track, and assess the key drivers, enablers and constraints, capabilities, and programs that will likely shape the future trajectories of China's military modernization – its pace, character, direction, magnitude, and impact on regional and global strategic environment. Notwithstanding contending assessments of the continuity and change in the Chinese science and technology base, its strengths and weaknesses, the key questions driving the debate are constant: Can China innovate and become an advanced technological power? What is the nature of China's efforts to transform its defense science, technology, and industrial capabilities? What are the critical sources of this innovation? What are the approaches and strategies that China is pursuing? How will China absorb, assimilate, and exploit its technological advancements for military purposes? And what are the key security ramifications for the region and the world? With the gradual, yet profound trajectory of China's rise as a greatest potential competitor to U.S. military predominance in the twenty-first century, Western strategists and policy analysts have demonstrated even greater need to assess the likely domestic, external, and technological drivers and aspirations shaping China's RMA-oriented military modernization. Indeed, the rising importance and scope of the debate has been tied to the substantive uncertainties in potential security challenges in the Asia Pacific region and the perennial need to minimize miscalculations and prevent strategic surprise.

References

Adams, Thomas K. 2008. *The Army After Next: The First Postindustrial Army.* Westport, CT: Praeger Security International.

Adamsky, Dima. 2008. "Through the Looking Glass: The Soviet Military-Technical Revolution and the American Revolution in Military Affairs." *The Journal of Strategic Studies* 31 (2): 257–294.

———. 2010. *The Culture of Military Innovation: The Impact of Cultural Factors on the Revolution in Military Affairs in Russia, the US, and Israel.* Palo Alto, CA: Stanford University Press.

Alberts, David, John Garstka, and Frederick Stein. 2000 [1999]. *Network Centric Warfare: Developing and Leveraging Information Superiority.* 2nd edition (Revised). Washington, DC: DoD C4ISR Cooperative Research Program.

Alberts, David, John Garstka, Richard Hayes, and David Signori. 2001. *Understanding Information Age Warfare.* Washington, DC: DoD C4ISR Cooperative Research Program.

Barnett, Thomas. 1999. "The Seven Deadly Sins of Network-Centric Warfare." *U.S. Naval Institute Proceedings* 125 (1): 1–13.

Biddle, Stephen. 1998. "The Past as Prologue: Assessing Theories of Future Warfare." *Security Studies* 8 (1): 1–74.

Bitzinger, Richard. 2006. *Transforming the U.S. Military: Implications for the Asia-Pacific.* Barton, Australia: Australian Strategic Policy Institute.

———. 2008. "The Revolution in Military Affairs and the Global Defence Industry: Reactions and Interactions." *Security Challenges* 4 (4): 1–12.

Blaker, James. 1997. *Understanding the Revolution in Military Affairs: A Guide to America's 21st Century.* Washington, DC: Progressive Policy Institute.

Blash, Edmund. 2003. "Network-Centric Warfare Requires a Closer Look." *Signal Magazine* (May Issue): 56–57.

Cebrowski, Arthur. 2003. *Military Transformation: A Strategic Approach.* Washington, DC: Office of the Secretary of Defense.

Cebrowski, Arthur, and John Garstka. 1998. "Network-Centric Warfare: Its Origin and Future." *U.S. Naval Institute Proceedings* 124 (1): 28–35.

CIA Directorate of Intelligence. 1989. "The Nature of Soviet Military Doctrine." *(SOV 89-10037CX).*

CIA National Foreign Assessment Center. 1981. "The Development of Soviet Military Power: Trends Since 1965 and Prospects for the 1980s." *(SR81-18935X).*

CIA Office of Soviet Analysis. 1984. "Soviet Ground Forces Trends." *(SOV 84-10173).*

———. 1985. "Trends and Developments in Warsaw Pact Theater Forces." *National Intelligence Estimate* (NIE 11-14-85).

———. 1987. "The Soviet Defense Industry: Coping with the Military-Technological Challenge." *(SOV 87-10035DX).*

Cohen, Eliot. 1996. "A Revolution in Warfare." *Foreign Affairs* 75 (2): 37–54.

Cooper, Jeffrey. 1994. "Another View of the Revolution in Military Affairs." In *In Athena's Camp: Preparing for Conflict in the Information Age,* edited by John Arquilla and David Ronfeldt, 99–139. Santa Monica, CA: RAND.

Cordesman, Anthony, and Paul Frederiksen. 2006. *Is Defense Transformation Affordable?* Washington, DC: Center for Strategic and International Studies.

Davis, Paul. 2001. *Effects-Based Operations: A Grand Challenge for the Analytical Community.* Santa Monica, CA: RAND.

Deptula, David. 2001a. *Effects-Based Operations: Change in the Nature of Warfare.* Arlington, VA: Aerospace Education Foundation.

——. 2001b. "Firing for Effects." *Air Force Magazine* 84 (4): 46–53.

Dombrowski, Peter, and Andrew Ross. 2008. "The Revolution in Military Affairs, Transformation and the Defense Industry." *Security Challenges* 4 (4): 13–38.

Fadok, David. 1995. *John Boyd and John Warden: Air Power's Quest for Strategic Paralysis.* Montgomery, AL: Air University Maxwell AFB School of Advanced Airpower Studies.

Farrell, Theo, Terry Terriff, and Frans Osinga. 2010. *A Transformation Gap? American Innovations and European Military Change.* Palo Alto, CA: Stanford University Press.

Freedman, Lawrence. 2006. *The Transformation of Strategic Affairs.* London, UK: International Institute of Strategic Studies.

Galdi, Theodor. 1995. *Revolution in Military Affairs? Competing Concepts, Organizational Responses, Outstanding Issues.* CRC Report for Congress (95–1170F). Washington DC: The Congressional Research Service.

Garstka, John, and Richard Hayes. 2003. *Power to the Edge: Command, Control in the Information Age.* Washington, DC: DoD C4ISR Cooperative Research Program.

Gray, Colin. 2006. *Strategy and History: Essays on Theory and Practice.* London, UK: Routledge.

Holland, William J. 2003. "What Really Lies Behind the Screen." *U.S. Naval Institute Proceedings* 129 (4): 73–75.

Knox, MacGregor, and Williamson Murray. 2001. *The Dynamics of Military Revolution 1300–2050.* Cambridge, UK: Cambridge University Press.

Krepinevich, Andrew. 1992. *The Military-Technical Revolution: A Preliminary Assessment.* Washington, DC: Center for Strategic and Budgetary Assessments.

——. 1994. "Cavalry to Computer: The Pattern of Military Revolutions." *National Interest* 37 (Fall): 30–43.

——. 2002. "Defense Transformation." Testimony for the United States Senate Committee on Armed Services (April 9), Washington, DC.

Libicki, Martin, and James Hazlett. 1994. *The Revolution in Military Affairs.* Prepared by CDR James Hazlett, and Dr. Martin Libicki, 1–4. Washington, DC: Institute for National Strategic Studies.

Marshall, Andrew. 1992. "Foreword." In *The Military-Technical Revolution: A Preliminary Assessment*, edited by Andrew Krepinevich. Washington, DC: Center for Strategic and Budgetary Assessments.

Mattis, James. 2008. "USJFCOM Commander's Guidance for Effects-Based Operations." *Parameters* 51 (Autumn): 18–25. Reprinted, *Joint Force Quarterly* (JFQ).

Mazarr, Michael. 1994. *The Revolution in Military Affairs: A Framework for Defense Planning.* Carlisle, PA: Strategic Studies Institute.

Mets, David. 1999. *The Air Campaign: John Warden and the Classical Airpower Theorists.* Montgomery, AL: Air University Maxwell AFB.

Metz, Steven. 2006. "America's Defense Transformation: A Conceptual and Political History." *Defense Studies* 6 (1): 1–25.

Metz, Steven, and James Kievit. 1995. *Strategy and the Revolution in Military Affairs: From Theory to Policy.* Carlisle, PA: Strategic Studies Institute.

Mitchell, Paul T. 2006. "Network Centric Warfare: Coalition Operations in the Age of US Military Primacy." *Adelphi Papers* (385): 1–71.
——. 2009. *Network-Centric Warfare and Coalition Operations: The New Military Operating System.* London, UK: Routledge.
Murray, Williamson. 1997. "Thinking About Revolutions in Military Affairs." *Joint Forces Quarterly* 15 (Summer): 63–70.
Naveh, Shimon. 1997. *In Pursuit of Military Excellence: The Evolution of Operational Theory.* London, UK: Frank Cass.
O'Rourke, Ronald. 2007. *Defense Transformation: Background and Oversight Issues for Congress.* CRC Report for Congress (RL32238). Washington, DC: Congressional Research Service.
Odeen, Philip, Richard Armitage, Richard Hearney, *et al.* 1997. *Transforming Defense: National Security in the 21st Century.* Washington, DC: National Defense Panel.
Office of Force Transformation (OFT). 2005. *The Implementation of Network-Centric Warfare.* Washington, DC: U.S. DoD.
Ogarkov, Nikolai. 1985. *Istoriya Uchit Bditelnosti (History Teaches Vigilance).* Moscow, Russia: Voenizdat.
Owens, William. 1996. "The Emerging U.S. System-of-Systems." *U.S. Naval Institute Proceedings* 63 (February): 36–39.
——. 2002. "The Once and Future Revolution in Military Affairs." *Joint Forces Quarterly* 31: 55–61.
Parker, Geoffrey. 1976. "Military Revolution 1560–1660 – A Myth?" *The Journal of Modern History* 48 (2): 196–214.
Petersen, Philip. 1984. "The Modernization of Soviet Armed Forces." *NATO's 16 Nations* 31 (4): 34.
Raynolds, Kevin. 2006. *Defense Transformation: To What? For What?* Carlisle, PA: Strategic Studies Institute.
Sanger, David. 2001. "Bush Details Plan to Focus Military on New Weaponry." *New York Times*, February 14, 1.
Shimko, Keith. 2010. *The Iraq Wars and America's Military Revolution.* Cambridge, UK: Cambridge University Press.
Smith, Edward. 2003. *Effects Based Operations: Applying Network-Centric Warfare in Peace, Crisis, and War.* Washington, DC: The Command and Control Research Program.
——. 2006. *Complexity, Networking, and Effects-Based Approaches.* Washington, DC: DoD C4ISR Cooperative Research Program.
Sullivan, Gordon. 1994. "Force XXI: Digitizing the Battlefield." *Army Research Development Acquisition Bulletin* (November–December): 2–4.
Sullivan, Gordon, and Anthony Coroalles. 1995. *The Army in the Information Age.* Carlisle, PA: Strategic Studies Institute.
Toffler, Alvin. 1984. *The Third Wave.* New York, NY: Morrow.
Toffler, Alvin, and Heidi Toffler. 1993. *War and Anti-War: Survival at the Dawn of the 21st Century.* Boston, MA: Little, Brown and Company.
U.S. Air Force. 2003. *The U.S. Air Force Transformation Flight Plan.* Washington, DC: Future Concepts and Transformation Division.
U.S. DoD. 1982. *Field Manual 100–5: Operations.* Washington, DC: Department of the Army.
——. 1996. *Joint Vision 2010.* Washington, DC: Office of the Joint Chiefs of Staff.
——. 2001. *Quadrennial Defense Review.* Washington, DC: U.S. DoD.
——. 2003. *Transformation Planning Guidance.* Washington, DC: U.S. DoD.

——. 2006. *Quadrennial Defense Review.* Washington, DC: U.S. DoD.

U.S. Department of the Navy. 2004. *Sea Power 21: Projecting Decisive Joint Capabilities.* Washington, DC: Department of the Navy.

U.S. Joint Forces Command. 2001. *Effects-Based Operations White Paper v.1.0.* Norfolk, VA: Concepts Department J9.

Vego, Milan E. 2006. "Effects-Based Operations: A Critique." *Joint Forces Quarterly* 41 (Spring): 51–57.

Watts, Barry. 1995. "What is the Revolution in Military Affairs?" *Northrop Grumman Analysis* 703 (351): 2.

3 Creating a reverse asymmetry

Military innovation concepts, issues, and debates in the IDF

Ever since the 1973 Yom Kippur War, Israel's national security environment has experienced gradual shifts resulting from a confluence of both external and internal factors that increasingly challenged Israel's national security consensus and essentially propelled changes in IDF's military commitments and operational conduct. The salience of traditional conventional threats facing Israel has relatively declined, while the asymmetric forms of warfare – WMD threats, proliferation of ballistic missile systems coupled with the increasing low-intensity conflicts (LICs) and non-linear threats – have increased, changing Israel's strategic assessments and patterns in the use of force. While the IDF has pioneered the IT-RMA-related capabilities such as network-centric warfare (NCW), stand-off precision strike, real-time battlefield intelligence, rapid target acquisition, integrated command and control, and remotely piloted vehicles, the IDF has viewed these innovations not as disruptive or revolutionary but rather as relevant adaptive measures that "fit new techniques, inventions, or operational outlooks into deeply rooted and relatively fixed military paradigm" (Cohen, Bacevich, and Eisenstadt 1998, 68). In time, however, the continuous adjustment in IDF's existing military means and methods led to multiple adaptations and bottom-up learning processes, which subsequently accelerated Israel's military innovation. As Dima Adamsky noted, "the technological seeds of Israeli RMA preceded the conceptual ones" (2010, 93).

It is in the context of these developments that the conceptual adoption and adaptation of the RMA-oriented military innovation within the IDF can be situated. First, the development of Israeli RMA concepts in the mid-1990s can be seen in *a broader framework of Israel's national security debate* on how to respond to the progressive complexity of security challenges – both from the current operational and long-term strategic perspectives. In this aspect, Israel's RMA discourse has reflected a continuous debate between the proponents of the traditional concept and those arguing for new military thinking within the IDF. The former camp argues that the increasing reliance on advanced weapon technologies will not ensure that Israel will use them effectively. In this line of thought, the political and geostrategic circumstances by which Israel's security dilemmas are bound remain largely intact – Israel cannot force its enemies to end the overall conflict through military means and decision alone. Moreover, the RMA as a theory has been largely irrelevant for Israel, as the IDF has used RMA concepts and technologies long before they have even presented as a paradigm shift in strategic discourse. Indeed, according to this school of thought, the IDF had a greater impact on the development of RMA concepts through its warfighting experience, rather than vice versa. Therefore, conceptualizing change within the IDF is not bound to the emergence of the RMA paradigm per se as in the Soviet or U.S. strategic thought, but can be attributed to the continuous search for relevant solutions and responses to Israel's changing strategic realities and unique operational needs. On the other hand, proponents of the RMA have emphasized that Israel's traditional security concepts are obsolete and need to be readjusted in response to asymmetric or "hybrid" forms of warfare. In this view, the RMA is not so much about acquiring military-technological capabilities, but developing relevant operational concepts and organizational structures that may effectively utilize these technologies to deal with the progressive complexity of security challenges facing Israel. That said, however, the IDF should take advantage of the "best technologies" offered by the RMA together with the "best of Israel's human capital" in order to ensure its qualitative military superiority in the twenty-first century. Ultimately, an RMA-oriented military innovation is imperative for Israel, as it faces persisting, imminent, and constantly evolving security challenges that require continuous adaptation. In other words, exploiting RMA-oriented technologies and concepts is not a question of choice but a necessity. Israel must use the best available technologies with innovative and adaptable operational concepts as force multipliers to offset its strategic inferiority and maintain its qualitative strategic edge over its enemies.

Following the lines of the above argument, this chapter is structured into two main parts. The first section presents the baseline of Israel's traditional security paradigm – its assumptions, foundations, and main elements, including deterrence, early warning, and rapid military decision. It also characterizes the traditional patterns in IDF's operational conduct and use of force – offensive ethos, preemption, and rapid maneuver. The second part of the chapter then identifies the sources of change in Israeli military thought and development of RMA-oriented operational concepts. In doing so, it examines four phases that shaped the Israeli RMA trajectory: (1) the Yom Kippur War; (2) the "offense-defense" debate and "saturated battlefield" concepts of the 1980s; (3) future battlefield concepts and the emergence of the systemic operational design (SOD) theory and methodology in the 1990s; and (4) the SOD debate and new concepts of operations debate from the late 2000s onwards.

Israel's traditional security concept

The diffusion of Israel's RMA-related concepts, processes, and debates can be situated in a broader context of change and continuity of Israel's *concept* of national security. Historically, Israel has refrained from publishing a formal document pertaining to a comprehensive national security strategy, periodic defense white papers, military doctrines, or similar documents issued by defense ministries of national security organs in many other countries (Heller 2000). It was only in spring 2006, for the first time in history, that Israel formulated its national security concept in written form, when the Committee Charged with Formulation of National Security Conception under Dan Meridor submitted its report (Adamsky 2010). Prior to that "Israel has developed only a national security concept but no national security theory and it has never produced a single formal document presenting the means and the ends of its national security policy" (Bar-Joseph 1998b, 147). This is because several inter-related factors have over time shaped Israel's defense policy making: (1) the need to maintain operational adaptability and decision-making flexibility in an unpredictable security environment; (2) the traditional nature of extensive secrecy on security matters in Israel; (3) persisting internal political considerations and security debates regarding force organization, resource allocation, and strategy and tactics; and (4) the necessity to focus on daily operational needs, issues, and problems that preclude theoretical and conceptual advancement. Hence, instead of a formal comprehensive doctrine, Israel has institutionalized a distinct set of principles – *national security conceptions* – that have subsequently guided the character and direction of Israel's defense strategies, operational use of force, and strategic culture, as reflected in the thinking of the country's political-military elite (both active and retired), as well as in the education and training of IDF officers (Tal 1977; Levite 1989).

According to Dan Horowitz, "all Israeli thinking on national security begins from the premise that Israel is engaged in a struggle for its very survival … given the perception that Israel is in a perpetual state of 'dormant war' even when no active hostilities exist" (1993, 1). Accordingly, Israel's traditional security concept is based on assumptions of a small state facing conditions of *geostrategic inferiority*, defined by Michael Handel (1973) as "basic, permanent strategic factors that characterized Israel's circumstances, defined threats and constraints, and determined the goals and aims that could be adopted within the limits of its power." These include Israel's geostrategic location, its small population, resource constraints, and superpower involvement in the region (Posen and Ben-Horin 1981, 5). First, Israel *lacks strategic depth*; the country is locked in a geographically small size with heavily concentrated population, industry, and infrastructure areas that are in close proximity to the country's borders and surrounded by adversaries (actual or potential) on three sides. From a strategic perspective, Israel has a dangerously short warning time for the interception or evasion of a potential, surprise enemy attack or incursion. All Israeli airfields are within three to five minutes' flying time from Syria and Jordan, as well as surface-to-surface missiles (SSMs) are capable of hitting any target in Israel from Egypt, Syria, Jordan, Iran, Iraq, and Saudi Arabia (Handel 1996, 538). Hence, in a time of war, Israel cannot trade space for time, nor can it afford to lose a single war (Cohen, Bacevich, and Eisenstadt 1998). As David Rodman noted, "this geographical situation early on led Israeli defense planners to the conclusion that Israel could not afford to 'host' either a full-scale war or a sustained low-intensity campaign on its territory. A sustained low-intensity campaign, they reasoned, would inevitably result in substantial damage to Israeli society, while a full-scale war could undermine the very survival of the state" (2001, 72). Consequently, geographical factors had a substantial impact on the trajectory of Israeli security conceptions, IDF operational doctrine of war, as well as on its organizational structure.

Second, *demographic inferiority* stemming from the population disparity between Israel and its neighbors served also as a key factor in shaping Israel's security perceptions. Leading Israeli strategists such as Yitzhak Ben-Israel often point to asymmetric population ratios, that is, Israel vs. its direct neighbors, 1/15; Israel vs. Arab world, 1/40; Israel vs. Moslem world, 1/200 (Ben-Israel 2009). Israel's demographic inferiority has been a particularly important factor during its formative years in the late 1940s, when Israel's Jewish population was only around 650,000 (Creveld 1998, 156). This created a broad psychological perception characterized by David Ben-Gurion as "the few against the many" (1953, 43). Throughout its history, Israel's numerical demographic inferiority has essentially imposed limitations on the maximum size of the armed forces it can mobilize while simultaneously forcing Israel to factor in a high degree of military participation ratio (MPR) – the proportion of general population in military service in order to offset the quantitative demographic superiority of Israel's adversaries. While Israel has over the years sustained one of the highest MPR in the world, it has been insufficient to alter the constant numerical inferiority vis-à-vis large Arab armies. As Barry Posen and Yoav Ben-Horin noted, "the manpower limitations render the Israelis in their own eyes particularly vulnerable to a major surprise attack on the one hand, and to extended strategies of attrition on the other" (1981, 6). Therefore, Israel had to rely on a three-tiered defense service system based on a relatively small number of career professionals and regular conscripts, including women in noncombat roles, with nearly 80 percent of the combat echelons comprised of well-trained, combat-ready reservists (Bar-Joseph 2004). The reliance on reservists, however, has exposed sociopolitical, economic constraints as well as military operational limitations. These include the limitation of the duration of any full-scale military operation, the high degree of defense burden caused by emergency call-ups or mobilization, disruption of the ordinary functioning of Israel's economy and society (Horowitz 1987). Ultimately, Israel's small population has amplified the country's psychological sensitivity and collective anxiety to the casualties of war, demonstrated on front pages of Israel's newspapers that often publish photographs and biographies of fallen soldiers. Over time, this prevailing sensitivity has shaped the conduct of wars, national security policies, as well as design of Israel's major weapon systems (Goodman and Carus 1990, 50).

Third, Israel's traditional security concept acknowledges a disparity of economic and natural resources relative to the surrounding Arab states – from shortage of water supplies to nearly complete external dependence on critical energy sources such as oil, which amplified the country's sense of vulnerability. According to Shai Feldman, "securing adequate supplies was another source of Israeli worries. Indeed, Israel's small population and industrial infrastructure made it especially sensitive to interference with its supply routes and concerned that the country's ability to obtain essential goods in time of war might be jeopardized" (1997, 10). Therefore, Israel cannot afford to sustain lengthy wars or protracted wars of attrition without outside material and political support. Ultimately, *direct and indirect involvement of superpowers* in the Middle East during the Cold War had significant implications on Israeli national security concept as well as on the conduct of Arab–Israeli wars. According to Avi Kober, as early as in the aftermath of the 1956 Sinai War, Israel concluded that the superpowers would never allow either party to achieve absolute decision on the battlefield, but would intervene at a certain point to freeze the war and initiate a ceasefire. From an Israeli viewpoint, such superpower intervention was a dual function of Israel's gains on the battlefield (the more extensive gains, the greater the chances for involvement) and time required for the superpower to reach consensus and decision (Kober 1995). Cases of direct superpower involvement included, for example, pressures by the United States not to take preemptive actions against a mounting threat (i.e., May–June 1967, October 1973), manipulation of arms supplies, denial of victory (i.e., August 1970, October 1973), Soviet participation in combat (1970) coupled with Soviet patronage of some Arab armies, and threats by the Soviet Union to limit Israeli war objectives in 1967 in Syria, in 1970 in Egypt, and in 1973 in both (Posen and Ben-Horin 1981, 7–8). On the other hand, indirect superpower involvement constraining Israel's strategic choices became the U.S. foreign assistance to Israel (averaging about US$2.1 billion annually in military financing and $800 million in economic aid), forcing Israel to acquire U.S.-made military technologies while restricting Israel's arms exports.

Israel's defense strategy: three pillars

The realities of Israel's external environment, internal resource constraints, and threat perceptions of its political-military elite have essentially shaped Israel's strategic and security conceptions, military thought, and operational responses. Since the early 1950s, Israeli leaders recognized that the sheer disparity between the Israeli and Arab territories, manpower, and resources prevents Israel from compelling Arab states to accept Israel's existence by military means alone (Bar-Joseph 2000). Israel, they believed, cannot achieve a "final victory" vis-à-vis its quantitatively superior Arab enemies. As David Ben-Gurion wrote in 1955 (Handel 1996),

> From our point of view, there can never be a final battle. We can never assume that we can deliver one final blow to the enemy that will be the last battle, after which there will never be the need for another clash or that the danger of war will be eliminated. The situation for our neighbors is opposite. They can assume that a final battle will be the last one, that they may be able to deliver such a blow to Israel that the Arab–Israeli problem will be liquidated.

Accordingly, at the grand strategic level, Israel's national security concept has been defensively oriented, but at the operational level it is offensive. As long as Israel is not able to end regional conflicts by political means, it aims at preventing escalation to a full-scale conventional war by deterring Arab states and their proxies and if deterrence fails to defeat them. Underscoring Israel's traditional national security concept are three mutually dependent strategic pillars: *deterrence, early warning, and rapid military decision* (Kober 1995; Bar-Joseph

1998a, 2000, 2004; Cohen, Bacevich, and Eisenstadt 1998). First, Israel has relied on a variety of deterrence strategies – using public or tacit military threats to influence strategic calculations of its enemies and persuading them that the cost or risks of their particular course of action might outweigh potential benefits (Inbar and Sandler 1994). Uri Bar-Joseph identifies at least four types of deterrence in Israel's strategic thought, categorized by specific objectives: (1) *current deterrence* – seeking to prevent the escalation of hostile Arab acts at the level of LICs; (2) *specific deterrence* – aimed at preventing Arab initiation of crises that threaten Israel's vital and peripheral interests; (3) *strategic deterrence* – to prevent a general large-scale war; and (4) *cumulative deterrence* – attempting to convince the Arab world that the Arab–Israeli conflict cannot be solved militarily in the long run at an acceptable cost (Bar-Joseph 1998a). Both strategic and specific deterrence in Israel's defense strategy has been also linked to the Israel's policies of "red lines" or *casus belli*, designed to prevent "serious deterioration in the state's strategic situation, which could result from certain incremental short-of-war Arab moves" (Posen and Ben-Horin 1981, 17). If crossed, *casus belli* signaled the failure of deterrence and increased the possibility of Israel's military response. Inherently, they also provided a basis for international justification of such military action (Cohen, Bacevich, and Eisenstadt 1998). Each deterrent type has subsequently affected the scope and intensity of Israel's military deployment, employment, and use of force. These can be conceptualized in two basic modes: *denial* and *punishment*. The former refers to both preventive and preemptive conventional military strikes in the context of both current and strategic deterrence to deny hostile Arab states success on the battlefield – that is, denying the Arab states the capabilities to carry out their hostile intentions. The latter, punitive element refers to the destruction of enemy's military forces and infrastructure, penetration of enemy lines and infliction of limited strategic damage beyond the battlefield, substantial destruction of enemy's arsenal, and capturing enemy territory for bargaining purposes in the context of specific deterrence.

Notwithstanding the prevailing conventional dimension of Israel's deterrence, the country has also relied on its undeclared "nuclear option" to amplify its deterrent posture. In particular, Israel has employed a policy of nuclear opacity (*Amimut*) – never admitting the possession of nuclear weapons, albeit not denying them either (Cohen 1998; Karpin 2006). On May 18, 1966, Israeli Prime Minister Levi Eshkol refined the Israeli conceptualization of the country's "nuclear policy" in the Knesset to a declaratory formula that has been subsequently adopted by all Israeli prime ministers and which has remained intact to this day: "Israel will not be the first to introduce nuclear weapons in the Middle East." In 1981, after the annihilation of the Iraqi nuclear reactor Osirak by Israeli warplanes, Israel proclaimed the so-called "Begin Doctrine" that "it would not allow itself to be the second country to introduce nuclear weapons in the Middle East." Referring to the role of "nuclear ambiguity" in Israel's defense strategy, Shimon Peres stated that "nuclear 'fog' was always part of Israel's national security concept" (Feldman 1997, 16). Israel's nuclear opacity coupled with the "bomb in the basement" policy – the absence of overt nuclear testing – has influenced other nation's perceptions, strategies, and action, including deterring conventional attacks by enemy states; deterring all levels of unconventional (chemical, biological/nuclear) attacks; preempting enemy state nuclear attacks; supporting conventional preemptions against enemy state nuclear assets to deter enemy's counterretaliation; and supporting conventional preemptions against enemy nonnuclear assets (Beres 1995). In this context, the underlying secrecy that Israel has maintained on its nuclear program has created much suspicion and speculation abroad that Israel indeed does have nuclear weapons (Rosen 1977). Furthermore, Israel's "bomb in the basement" policy has essentially enabled the country to successfully navigate through the conundrums of the Cold War's nuclear age, deterring Soviet intervention in Arab/Israeli wars as well as providing an "equalizing" insurance policy of last resort vis-à-vis surrounding, quantitatively superior Arab armies. At the same time, Israel was able to evade the nuances of international arms control treaties (NPT), sanctions, and inspections by the International Atomic Energy Agency designed to prevent the proliferation of nuclear weapons. Israel's deliberate nuclear ambiguity may have also played a role in strategic calculations of Egypt (1979) and Jordan (1994) to conclude peace treaties with Israel.

The second pillar of Israel's national defense strategy, closely associated with deterrence, has centered on the need for *early strategic and tactical warning.* According to Israel Tal, the concept of early warning in the Israeli security doctrine has never been considered as "warning of intentions" but "as the formation of a threat as an objective reality … [that] does not depend on information or appreciations concerning a priori intensions but rather assessments of the feasibility of actions, based on information and the development of a threat on the ground that are likely to indicate enemy intentions to go to war with Israel" (2000, 78). In the words of Kober, "[early warning has aimed] at alerting policy-makers to the possibility that deterrence is in effect failing and that war might be in the offing" (1995, 189). Hence, both strategic and tactical intelligence capabilities have been perceived as essential – *a first line of defense* – providing Israel's policy makers with a timely strategic warning on any developments that might endanger Israel's national security. In this context, Israel is the only country in the world that places the status of its intelligence branch to that of the air force, army, and navy (Handel 1996, 541). Throughout its history Israel has been relatively more successful in providing high-level intelligence at the operational and tactical levels rather than estimating the emergence of strategic threats and opportunities (Bar-Joseph 2007). Most notable example of this pattern is the 1973 Yom Kippur War, when Israel's military intelligence (AMAN) failed to interpret strategic warnings of an impending coordinated attack by Syria and Egypt, but were able to provide critical information on the tactical and operational levels. Similarly, Israel's military intelligence failed to both provide strategic warning prior to the outbreak of the first Intifada (1987) and estimate the magnitude of the second Intifada (2000). However, at the operational level, "it provided excellent operational intelligence that enabled Israel's security services to prevent numerous planned terrorist acts and hit many terrorist in targeted killings that, in most cased, involved limited collateral damage" (Bar-Joseph 2007, 584). In either case, Israel has over the years invested extensively in an array of sophisticated technological and electronic means – from ground and airborne to space-based platforms for collecting, analyzing, and disseminating visual as well as human intelligence information pertinent to Israel's national security (Bar-Joseph 1998a).

Finally, if both deterrence and strategic warnings failed, the third pillar of Israel's national defense strategy emphasized *rapid military decision* through "decisive" victory on the battlefield. Decisiveness in battle has been viewed by Israeli political and military leadership as "compelling the enemy to accept a cease-fire on terms which are favorable to us" (Dayan 1999, 4). Specifically, according to 1998 *IDF Dictionary of Concepts*, "the break of the enemy's resistance power to act effectively against us by creating a situation in which (according to the decision-maker's estimate) the conditions necessary to attain the set goal [of the war] are achieved. A state of decision is recognized as such when the enemy loses its ability to act effectively against us" (Bar-Joseph 2004, 141). In other words, Israeli strategists understand that given the inability to achieve strategic or political victory amid the prevailing territorial, demographic, and foreign policy constraints, ensuring a successful deterrent posture is also difficult if not impossible to achieve. Therefore, Israeli deterrence has been limited to short-term successes and outcomes of particular "military rounds" (i.e., Six-Day War, Yom Kippur War, Lebanon War I and II). In each "round," however, Israel aimed at achieving a "decisive operational victory" (*hachra'a*) in order to extract a painful price from its enemies and to ensure that they can no longer carry the fight in the particular military "round" (Inbar and Sandler 1994). As Ariel Levite noted, "only by inflicting such a humiliating defeat (in terms of casualties, destruction, and loss of territory) or succession of defeats, it was maintained, could the Arabs be dissuaded, temporarily or permanently, from launching a war against Israel" (1989, 43).

In theory, rapid military decision would leverage the failure of deterrence, significantly prolong the gap between military rounds, substantially weaken military capabilities of Israel's adversaries and their efforts to sustain their attrition strategies, and, in the long term, possibly lead to peaceful resolution of the Arab–Israeli conflict. Therefore, rapid military decision became the central, if not the most important, component of Israel's security concept, signifying both Israel's strategic objectives as well as operational means to achieve them. It stipulated the necessity to achieve a swift battlefield victory in order to minimize military and civilian casualties, mitigate the costs of prolonged wars, strengthen national consensus supporting military action, decrease the magnitude of superpower involvement, and eventually return the country back to normalcy. In doing so, it provided an essential element, if not always sufficient in itself, to attain a favorable political bargaining position after the war (Posen and Ben-Horin 1981, 18). Ultimately, rapid military decision increased the credibility of Israel's willingness to use its military power and thus the credibility of deterrence. As Shimon Peres noted, "Deterrent power – the power to deter the enemy from attacking – is also the power to vanquish him if he should fail to be deterred" (1970, 218).

Patterns in IDF's operational conduct

Intrinsically, "the IDF is the 'Israel *Defense* Forces' by appellation but 'Israel *Offense* Force' in substance" (Tal 2000, 43). In order to attain a quick, decisive, and visible victory on the battlefield and achieve a continual military advantage over its enemies, the IDF has developed flexible operational conduct that employed offensive maneuver-style warfare, often deep inside the enemy territory. Given Israel's geostrategic, demographic, economic, and diplomatic constraints, offensive maneuver warfare has been viewed as "the only strategic defense option available to the IDF" (Posen and Ben-Horin 1981, 30) – that would minimize casualties, shorten the duration and costs of conflict, and preclude superpower intervention, while at the same time strengthen Israel's deterrent credibility by delivering a decisive blow to the quantitatively superior, albeit organizationally inferior Arab armies. In this context, Israel's operational conduct has inherently aimed at creating a "reverse asymmetry" – exploiting the IDF's qualitative superiority in order to offset its quantitative inferiority. Whenever deterrence failed or when Israel's basic security was endangered, the IDF would take the initiative to conduct high-intensity maneuver warfare, creating artificial strategic depth by transferring war into enemy territory, flanking and outmaneuvering its enemies to reduce casualties and achieve tactical surprise, while implementing operational and command flexibility that allowed more efficient adaptability to the changing circumstances.

According to Shimon Naveh, IDF's operational conduct can be summarized thus: "an offensive ethos of a preemptive attack executed by a combination of armor-heavy and air striking forces deep into enemy territory, [that] would certainly bring about the destruction of its military force and thus remove the immediate strategic threat to Israel" (1996, 169). The IDF developed its organizational force structure, training, and force employment on three mutually dependent tenets: (1) *offense*, (2) *preemption*, and (3) *rapid maneuver.* First, the principle of offense has been associated with a swift and resolute transfer of war into the enemy territory as soon as hostilities break out with the aims to offset Israel's lack of strategic depth and overall numerical inferiority, retain operational control and initiative on the pace and intensity of combat, disrupt or deny the enemy's plans, shorten the duration of conflict, significantly reduce war costs, and ultimately strengthen Israel's deterrence by both denial and punishment (Tal 2000, 70). Ben-Horin and Posen clarify, "by seizing the operational initiative ... the IDF would concentrate forces at chosen points, attain local parity or even superiority, and seek decisive operational victory by swift disruption or destruction of enemy forces at critical junctures" (1981, 29).

Transferring war into enemy territory has been deeply embedded in Israel's strategic culture since the formation of the state and has endured in the IDF's operational conduct as "ironclad law" throughout the decades. Numerous references point to its lasting character and prevailing attitudes. One of its earliest references point to 1948, when Ben-Gurion noted, "if we are attacked and war is again forced on us, we shall not adopt a defensive strategy, rather we will move to an attack on the enemy – and as far as possible, in enemy territory … If they attack us as they did this time – we shall transfer the war to the gates of their country" (Levite 1989, 49). Similarly, in the early 1950s, Yigael Yadin, IDF's second chief of general staff (CGS), argued that "we cannot build a defense plan based on fighting in our own territory. That means, of course, that we have to transfer the war to the enemy's country … The operation goal of our planning is to take quick decisive action and to transfer the war to the enemy's territory … If we are attacked we should defend ourselves where attacked, but we have to attack at the point that we find most convenient" (Handel 1973, 34). More than three decades later, in 1985, Yitzhak Rabin, then Israel's defense minister, repeated the principle (Levite 1989, 53):

> Israel's basic security policy is … in the face of the existential threat posed by the Arab armies, our goal is first and foremost to deter from war. If our deterrent capability is insufficient, a decision must be gained in the most convincing manner – by doing maximum damage to the forces of the state or the states that attack us, while achieving maximum territorial gains – in order to bring about a ceasefire, at the request of the other side, with maximum speed and under conditions convenient for the State of Israel. What this means in practical terms is that in the event that deterrence proves insufficient, *we will have to transfer the war to enemy territory*. The import is that our security concept is basically defensive, but is applied by the IDF, whose main thrust is offensive.

Closely tied to the IDF's principle of offense has been the emphasis on preemptive and preventive types of wars. The former has aimed at deflecting an imminent threat, while the latter was designed to forestall deterioration in the balance of power (Kober 1995, 199). From 1948–1955, Israeli military doctrine did not endorse the principle of war initiation as the threat posed by the Arab armies was not perceived to warrant it, or the external and internal political circumstances coupled with IDF's limited war-fighting capability at that time seemed to preclude successful implementation of such doctrine (Levite 1989, 53). However, in the wake of the 1956 Sinai Campaign and gradual improvements in the IDF's operational capabilities, Israel's operational military doctrine began to increasingly shift toward patterns of "preemptive counteroffensive" or "anticipatory counterattacks" – the idea of striking first in the face of a major imminent threat (Allon 1969, 69–76). Through preemption, such as in the 1967 Six-Day War, the IDF could gain a number of military advantages that improve the probability of a decisive operation victory. In theory, by preemptive wars the IDF could achieve strategic surprise by immediately transferring the battle to enemy territory, denying the enemy its predetermined battle plans, preventing or disrupting the mobilization and concentration of enemy forces, shortening the duration of war, and, in doing so, mitigating the scope of potential superpower intervention. Furthermore, preemption would (at least theoretically) reinforce the credibility of Israel's deterrence strategy – especially its policies of "red lines" such as the use of force vis-à-vis concentration of Arab forces in the proximity of Israel's borders. However, it should be noted that amid a range of both external and internal political considerations, preemptive or preventive modes of warfare have not always represented automatic course of action by the IDF – as in the case of the "ironclad law" of transferring war into the enemy territory (Posen and Ben-Horin 1981, 34; Levite 1989, 58). For example, in 1960, Egypt concentrated the bulk of its offensive forces in the Sinai (Operation "Rotem"). Israel countermobilized; however, it did not strike. The crisis was diffused quietly with both armies pulling back to their original positions. Similarly, on October 6, 1973, on the eve of the Yom Kippur War, Israel was reluctant to initiate a preemptive strike vis-à-vis advancing Syrian and Egyptian armies, which has been later perceived by senior IDF officers as a serious miscalculation (Levite 1989, 59). In the early 1980s, during Ariel Sharon's tenure as defense minister, military preemption and prevention regained their standing in Israel's military doctrine – culminating in Israel's attack on Iraqi nuclear reactor Osirak at Tuwaitha in 1981 and the invasion of Lebanon in 1982. However, with a narrowing technological gap coupled with the changing threat spectrum brought by the introduction of tactical and intermediate-range ballistic missiles (such as the Soviet-made SS-21 in the Arab arsenals) and their WMD ambitions, Israel has been more reluctant to use preemptive or preventive attacks, which have become increasingly difficult to implement, and instead opted for a more defensive posture. However, preemptive or preventive strikes have not been entirely discarded as a possible option for particular politico-strategic circumstances (Inbar 1998, 72).

The third defining characteristic in IDF's operational conduct has been the emphasis on *rapid maneuver warfare* and its application across a broad spectrum of operations and environmental conditions. The IDF has defined maneuver as "the combination of force movement and fires with the objective of gaining an advantage over the enemy" (Glenn 2008, 9). From this perspective, maneuver has been conceptualized as a principle of war that aims to place the enemy in a disadvantageous position through a dynamic, flexible, and comprehensive application of combat power. In the Israeli case, this has meant exploiting three key elements at the operational and tactical levels: (1) *speed* in the context of combined arms warfare, (2) *surprise* through indirect approach, and, perhaps most importantly, (3) *flexibility* in command, plans, and operations. In particular, depending on battlefield conditions and situations, the IDF in combat concentrated different types of combined arms and units to maximize their effectiveness and employed them in fast-paced military operations that often used the line of least resistance or the line of least expectation – that is, bypassing frontal assaults and attrition battles, while encouraging Israeli soldiers to find operational methods that would achieve important objectives and reduce casualties (Creveld 1998; Rodman 2002). Taken together, operational mobility and maneuverability coupled with substantial firepower enabled the IDF to gain the initiative, shorten the duration of conflict, reduce casualties, forestall Arab mobilization of their superior quantitative resources, and preclude the anticipated superpower intervention in the conflict (Posen and Ben-Horin 1981).

Offensive maneuver warfare had implications on the IDF force structure, command and control, resource allocation and weapons procurement, training, and logistics. While in its early wars (1948–1956), the IDF relied predominantly on its mechanized infantry units supported by armored units and limited air power; from the 1960s onwards, the IDF's operational doctrine shifted toward combined arms warfare based primarily on heavy armored formations coupled with the use of air power. Israel's weapon of choice became the tank – its firepower and mobility coupled with its "armored shock" doctrine aimed at achieving rapid breakthroughs on all fronts and subsequently maneuver deep inside the enemy territory (Rodman 2002). In this context, tanks provided the IDF "a multi-purpose and versatile weapons system, usable for breaching enemy formations and fortifications, usable for both defense and offense, for armor battles, and for exploitation of success and pursuit; firepower and considerable mobility capabilities" (Kober 1995, 202). In order to use tanks, however, control of the air became a necessary condition – a role for the Israeli Air Force (IAF). Historically, the IAF has played a vital role in supporting IDF's offensive maneuver warfare with its multirole application of "air supremacy" and "air superiority," which have crystallized during the 1950s and 1960s. The former has referred to a continuous effort to achieve and maintain a clear-cut advantage in all dimensions concerning air power – that is, a qualitative edge in human resources, technology, and operational capability. The latter shaped and defined the course of a conflict – that is, air defense, air-to-air combat, battlefield interdiction, and strategic missions vital to Israel's national security (Dayan 1999).

Israel's RMA trajectory and strategic debates

For more than 50 years, the fundamentals of Israel's traditional security concept – a defensive strategy, executed offensively – have endured largely unchanged notwithstanding the persistent and often conflicting internal debates and assessments of Israel's national security environment. Its basic aim has been to deter aggression and, if deterrence fails, to ensure a rapid decisive victory in every military encounter by means of self-sufficient military capabilities. While some of its elements were conceptualized early in Israel's history as a response to a broad range of prevailing geostrategic constraints, others have evolved in the following decades as a result of lessons learned in Israel's direct military experience and changing politico-military circumstances. In this context, Ben-Horin and Posen concluded that "doctrinally, [the IDF] changed but little between 1967 and 1973, and appears to have changed in only an evolutionary way since then. Neither experience nor new technology have produced any strategic revolutions" (1981, 48).

However, following the 1973 Yom Kippur War, Israel's national security environment began to experience gradual shifts resulting from a confluence of both external and internal factors that increasingly challenged Israel's national security consensus and essentially propelled changes in IDF's military commitments and operational conduct. Notwithstanding the many lessons learned that have shaped IDF's military doctrine since its inception, none of the previous adjustments have questioned the efficacy of the traditional concept itself or the operational needs and performance of the IDF. The emergence of contending debates on how to retain Israel's strategic edge coupled with changes in Israel's strategic orientation in the 1970s, 1980s, and 1990s provides an important background for understanding the Israeli RMA debates in the twenty-first century. In particular, the conceptual adaptation of RMA from the mid-1990s can be positioned in the continuing strategic debate within Israel's political and military establishment on how to best respond to the progressive complexity of security challenges and threats. As this chapter argues, the IDF adopted selected RMA concepts in the late 1990s as a response to new strategic realities and operational demands rather than the emergence of the RMA paradigm per se.

Intellectual roots: the Yom Kippur War

The 1973 Yom Kippur War, also known as the October War, was a turning point in the way Israeli society and military viewed war, notwithstanding the concluding Israeli victory over Egyptian and Syrian forces, "Israel did not feel victorious" (Inbar 1983, 36). According to Martin van Creveld, "the October War was by far the most difficult and costliest ever fought by the IDF" (1998, 243). Similarly, Uri-Bar Joseph notes, "despite the fact the war ended with a limited Israeli victory on both front, the way it began, Israel's heavy casualties, and the IDF's inability to win a decisive victory at its end, have turned into the gravest event in the history of the country" (2008, 76). The Yom Kippur War had a number of significant strategic implications. Externally, it diminished Israel's deterrent value and the "myth of Israeli invincibility" in the Arab world, which viewed the October 1973 war as a victory (Inbar and Sandler 1994). The war also exposed Israel's dependence on external political, military, and economic aid, particularly from the United States, which provided a massive airlift of arms supplies to Israel in the later stages of the war. The combination of U.S. financial and military aid, coupled with diplomatic pressures on both Israel and Egypt in the years following the war, has inherently shaped the direction of the Egyptian–Israeli peace treaty negotiations and its signing in 1979 (Inbar 1983).

Internally, the war sparked a number of security debates on the events and consequences surrounding the war; IDF's operational and organizational conduct; and subsequently the necessity to enhance Israel's defense planning, military doctrine, training, and organizational force structure. Perhaps more importantly, it imposed much deeper self-reflection within the Israeli military establishment on the limitations of power and deterrence and the price of confining security to military operational dimensions (Bar-Joseph 2004). While no substantial disagreements emerged on the underlying logic behind Israel's strategy during the war, (1) denying the enemy's military gains and subjecting him to military defeat, (2) destruction of enemy forces and military infrastructure, and (3) ending the war on more favorable military and political terms prior to superpower involvement, its lessons exposed many flaws in Israel's early warning system, operational conduct, and doctrine. Attempting to draw appropriate lessons of the war, the official

Agranat Commission published its findings (to a limited extent) in early 1975. The report noted the personal and structural failures of Israel's strategic intelligence community and early warning systems and questioned traditional tenets and assumptions behind the IDF military doctrine. As Emanuel Wald noted, "the entire inventory of IDF concepts and myths disintegrated in the Yom Kippur War" (1992, 112).

Notable among the failures was Israel's armored (all-tank), rapid maneuver doctrine, designed to "absorb" and "brake" an Arab first strike and subsequently deliver a decisive counterattack through frontal "armored shock" supported by the air and ground forces. The traditional Israeli "all tank" method of armored warfare relied on relatively independent tank battalions – with their own mobility, forward movement, and accurate firepower to breach enemy defences (Luttwak and Horowitz 1975). This method emerged from the IDF experiences during the 1956 Suez Conflict and 1967 Six-Day War, which gave the impression that "wars on the ground are won by armor and armor alone" (Bolia 2004, 51). In the opening stages of the October War, however, the Arab armies utilized both direct and indirect means to offset Israel's "all tank" doctrine, including integrated, multi-layered array of defensive lines, replete with anti-tank barriers, fortifications, and obstacles, augmented by large-scale air defense systems in order to reduce the maneuverability of Israeli armor on land and freedom in the air (Levite 1989). In the Sinai, the Egyptian mechanized and armored columns were able to cross the Suez Canal, covered by dense layers of latest Soviet-made surface-to-air missile (SAM) systems, heavy artillery barrage, and infantry equipped with portable Sagger anti-tank guided missiles (ATGMs). Syrian tanks were able to penetrate Israeli defenses in the southern sector of the Golan front and threatened to move across the Jordan River into Israel's pre-1967 territory. The relatively well-balanced Egyptian/Syrian invasion plan thus mitigated, at least in the initial stages of the war, the effectiveness of Israel's tactical superiority in both air and ground combat (Luttwak and Horowitz 1975). It took the IDF several days to recover from the initial shock of surprise and to stabilize the lines on both fronts. In the weeks ahead, however, the IDF delivered decisive counterattacks by effectively modifying its "all-tank" approach into more flexible combined-arms operations – Israeli tanks now operated in an "artillery box" supported by mechanized and parachute infantry and the air force (Rodman 2002). For the first time, the Israeli Navy also utilized advanced antiship missiles and missile boats against a range of military and industrial targets on the Egyptian and Syrian coastal areas (Tzalel 2000). On October 24, 1973, the day of the cease-fire agreement, the IDF on the southern front led by Gen. Ariel Sharon stopped only about 60 miles from Cairo.

In retrospect, the Yom Kippur War caused tremors in the Israeli security establishment. The IDF realized that its enemies managed to leverage the IDF's early warning, armored combat and air superiority concepts and weapons. As a result, the IDF drew four important lessons: (1) the modern battlefield environment characterized by destructive firepower, accuracy of weapons, speed, and higher attrition rates is more lethal than ever before; (2) fighting demands highly trained and integrated forces – the successful integration of air power with ground forces will make a difference between victory or defeat; (3) tactical training makes a difference between success and failure; and (4) maintaining credible early warning and intelligence is essential to avoid strategic surprise. Subsequent studies of the 1973 Yom Kippur War shaped RMA visions for future warfare and defense system requirements while accelerating development and integration of new military technologies, including command, control, communications, and intelligence (C3I) platforms and precision munitions. As Adamsky noted, "after the [war], the IDF intensified their investment in PGM weaponry, sophisticated 'over-horizon' intelligence capabilities, and command and control systems" (2010, 4). The war also drew the interest of the U.S. military, which sent its strategists to Israel to exchange ideas, share lessons, and formulate the basis for a new operational doctrine that later formed into the ALB (Tomes 2004). Taken together, the Yom Kippur War served as the "single most important factor that has shaped the country's national security doctrine [until today]" (Bar-Joseph 2008, 70).

1980s: the "integrated battle" and offense-defense debate

During 1973 and 1985, the IDF experienced a dramatic infusion of modern weapon systems and technologies – the biggest and most expensive force built up in Israeli history. The number of main battle tanks (MBTs) increased from 1,900 to 3,600; armored fighting vehicles (AFVs) from 2,500 to 8,000; artillery pieces tripled to 2,000. The IAF's first-line combat aircraft grew from 400 to 700 and included a new generation of U.S.-built F-15 and F-16 fighters, armed with PGMs and supported by airborne electronic warfare (AEW) platforms. The total IDF force, including reservists, increased from 300,000 soldiers in 1973 to about 540,000 in 1982 (Creveld 1998, 252–53). The military build-up was designed to enhance Israel's defensive strategic posture to maintain balance of forces vis-à-vis Arab armies and have more IDF formations available for maneuver on two fronts simultaneously (Inbar 1983). Doctrinally, while there were no significant changes in the broader "absorb" and "break" concept followed by a "rapid maneuver," the IDF jettisoned the "all tank" doctrine and focused more on mobility and firepower capabilities in a relatively flexible combined-arms mode (Rothenberg 1979).

However, following the Egypt–Israel peace treaty in 1979 and the Lebanon War in 1982, the contours of Israel's security environment began to shift. Israeli military planners at that time focused predominantly on the possibility of a new war with Syria as well as on new modes of conflicts in urban areas (Cohen, Bacevich, and Eisenstadt 1998, 83). Syria after the Lebanon War pursued a policy of "strategic parity" with Israel, in order to "build up enough independent military power to pose a credible counter-weight to Israel without having to rely on an Arab coalition … for defensive purposes and perhaps even to undertake limited offensives" (Heller 2000, 22). The quantitative superiority of the Syrian army, increasingly amplified by new types of asymmetrical threats (SSMs plus WMDs) and low-intensity threats in urban areas, prompted a debate within the IDF on the future character of the battlefield and required doctrinal, organizational, and operational adjustments. The resulting "offense-defense" debate polarized Israel's military establishment. In particular, proponents of the traditional view – "tanks lead the way" – such as Israel Tal viewed Israel's new strategic position and operational problems as reinforcing the need for preemptive offensive-style maneuver warfare in the air, land, and sea. In this view, the IDF should exploit its superiority in battle mobility and flanking movements, seize the initiative and take the war into the enemy territory – a concept which has essentially stood the test of time (Tal 1989, 47). Any shift toward a defensively oriented doctrine would inherently mitigate Israel's deterrence capabilities while providing advantages for Arab attrition strategies in possible future conflicts.

On the other hand, contending views by reformers such as Ariel Levite argued that offensive actions in the emerging "saturated battlefield" – a dense array of enemy (Syrian) defensive fortifications and weapon systems aligned in greater depth – substantially limited the opportunities and applicability of traditional offensive maneuver and breakthrough battles, increasing the risks of high casualty rates and early superpower intervention. Accordingly, acquiring capabilities for long-range, precision firepower in-depth of the saturated battlefield, that is, using air force, may be more relevant and effective than the capability to maneuver (Rodman 2000). In other words, the concentration of precision firepower may become more important than concentration of forces (Kober 2001). The IDF should, therefore, develop more defensive-oriented concepts of operations and exploit advances in weapons technologies and systems – that is, precision munitions, automated command-and-control systems, and new generation of target acquisition capabilities – in order to create new war-fighting options for the IDF (Cohen, Bacevich, and Eisenstadt 1998, 84).

The offense-defense debate brought a disparity of views within Israel's military establishment – not only about the validity and relevance of existing operational concepts but also resulting organizational structures and force composition. The structure of the IDF in the early 1980s was still organized around armored formations to conduct high-intensity wars. However, the IDF faced increasingly more complex "combat zones" with amorphous nonstate actors assimilating in urban areas that precluded achieving rapid military decision and clear definition of operational objectives (Tamari 2001). These issues were raised in an internal report written in 1987 by Col. Emmanuel Wald. The "Wald Report" report was highly critical of the IDF's increasingly cumbersome and feuding bureaucratic force structure, along with anachronistic concepts of operations, which essentially led the IDF to failures in the 1982 War in Lebanon (Cordesman 2002, 198). At that time, the IDF denied and rejected the report, yet subsequent statements and reforms by senior IDF commanders in the early 1990s indirectly validated Wald's findings (Creveld 1998, 260). Subsequently, the IDF under Defense Minister Yitzhak Rabin considered both traditional and reformist arguments and essentially compromised by procuring advanced weapons systems and stand-off PGMs, but integrating them into the existing force structures and operational concepts (Adamsky 2010, 97).

Parallel with the introduction of new generation of weapons technologies and systems from the early 1980s, the IDF also began to develop innovative *tactical solutions* designed to offset new threats and deficiencies, primarily those exposed during the Yom Kippur War. In particular, the IDF focused on developing "integrated" combined arms capabilities that relied on force integration and force multiplication factors. These included close coordination and integration of air and ground forces, improved intelligence and early warning, streamlining of command control procedures, combined planning and training, and precision fire. During the mid-1980s, the IDF established a unified Ground Corps Command (GCC) tasked with doctrinal and operational integration, weapons acquisition, and combined training to achieve maximum coordination of ground elements. The IDF viewed the establishment of the GCC as essential to the process of modernization and adaptation to the modern high-tech battlefield (Abramowitz 1988). Modern warfare for the IDF at that time was conceptualized under the umbrella of "integrated battle." Maj. Gen. (ret.) Uri Saguy summarized the implementation of the concept: "The integrated concept is found here at every level…our goal is to arrive at equilibrium between the different IDF elements. Integrating our forces ensures that the tactics of each element are not isolated. We analyze the battlefield together, and then develop the strategy and means that produce the desired result" (Ben-Dor 1989, 7). Similarly, Maj. Gen. Amir Drori defined the IDF's integrated approach as "the exploitation of the combined potential of all formations, used with the intention of defeating the enemy on the battlefield [and] adapted to the task at hand" (Weinraub 1988a, 16).

The IAF also began developing its own "integrated" combat doctrine for the modern battlefield under the umbrella of "coordinated participation" (Weinraub 1988b). This meant IAF's integration with intelligence and ground-based elements in four areas: (1) direct close support; (2) coordinated participation, in which aircrafts are allocated a sector by the ground forces and allowed to operate with relative freedom; (3) interdiction; and (4) impending enemy command and planning capabilities. A notable case study of "coordinated participation" is the "Bekaa Valley Air War." In June 1982, the Syrian army deployed a dense belt of mobile SAM batteries, SA-6, throughout the Lebanon's Bekaa Valley. These batteries posed a direct threat to the IAF operating west of the valley as part of the operation "Peace for Galilee" (Grant 2002). Although Israel proclaimed a desire to neither engage the Syrian forces based in Lebanon nor initiate any war with Syria, a decision was made for a preemptive strike on these SAM sites (Zisser 2004, 202). Under an innovative tactical plan "Operation Mole Cricket 19" (*Mivtza Artzav Tsha-Esreh*), the IAF would first deploy unmanned drones or remotely piloted vehicles (RVPs) as decoys to "illuminate" and "fingerprint" enemy radar signals while collecting real-time intelligence and visual reconnaissance of the SAM sites (Sanders 2002). The Syrians launched most of their available missiles at the incoming drones, mistaking them as attacking Israeli aircraft. Meanwhile, as the Syrians reloaded their missiles, the radar frequencies from the drones were relayed to the IDF ground and airborne command posts, which coordinated AEW platforms to jam them and subsequently guided the IAF fighter-bombers armed with anti-radiation missiles and laser-guided bombs to attack the missile sites. In just 15 minutes, the IAF with its F-4 Phantoms destroyed 17 of the 19 SAM sites, with no losses (Grant 2002). In response, the Syrian air force scrambled its fighters (MIG-21s and MIG-23s) to intercept and repel the IAF jets – an encounter that turned into one of the largest air battles since the Korean War. The IAF had a decisive advantage with its real-time C3I capabilities delivered by the E-2C airborne warning and control system (AWACS). It also used the same RPVs flying in high altitudes over the Syrian airfields to monitor their take-off activities. With this intelligence, the controllers aboard the Israeli E-2C could then vector in formations of Israeli F-15s and F-16s into the engagement zones to intercept them. Simultaneously, the Syrian fighters were effectively cut off and blinded from their ground controllers by Israel's electronic warfare systems that jammed the Syrian C3 airwaves. Using advanced heat-seeking (AIM-9L) and radar-guided (AIM-7E/F) air-to-air missiles, the IAF then shot down 26 MIGs within 30 minutes. By the next day, the IAF pilots shot down about 82 Syrian fighter jets without losing any in air combat (Clary 1988).

Notwithstanding its limited scope and unique circumstances, the Bekaa Valley Air War essentially anteceded U.S. and Soviet RMA concepts with its images of network-centric electronic warfare, precision bombing, unmanned aerial vehicles (UAVs) in combat, real-time C3I, joint operations, and integrated air power (Adamsky 2010). It was viewed as a victory of the Israeli system over the Syrian (and Soviet) system (Schiff and Ya'ari 1984, 167). The Arab strategic paradigm relying on massive Soviet-built air defense networks of SAMs, which denied the freedom of action of the IAF during and after the Yom Kippur War, was no longer relevant. This propelled Israel's opponents to reevaluate their strategies, correct deficiencies, and seek alternative asymmetric means to counter the IDF's qualitative technological edge and superior manpower quality. At the strategic level, Israel's main opponents – Syria and Iraq – began to acquire asymmetric capabilities, that is, SSMs, while expanding their conventional arsenals. As David Ivry, former IAF Commander, noted, "that's when they started to buy the Scuds" (Grant 2002, 62). In retrospective, the Bekaa Valley experience did not translate into a notable paradigm shift for the IDF and its concepts or operations. Rather it reflected a traditional "relational" pattern of tactical and operational innovation aimed at retaining Israel's strategic edge in constantly changing operational conditions. As Adamsky noted, "Israel regarded the Bekaa battle as a successful synthesis of advanced technology and creative operational improvisation – the essence of its qualitative edge" (2010, 96).

1990s: future battlefield concepts

Historically, Israel has broadly distinguished two types of security threats: "basic or fundamental security" (*bitachon yisodi*) and "current security" (*bitachon shotef*, or in short *batash*). The former has been referred to major conventional wars – real and potential that stipulated major risks for Israel's existence; the latter represented LICs, terrorist threats and attacks, border skirmishes, and enemy intrusions that harmed but did not seriously threaten the existence of Israel. The prioritization of basic security, which has historically transcended all differences in ideology and politics, can be seen at Israel's core basic security concept (Cohen, Bacevich, and Eisenstadt 1998). In this context, Israel has also traditionally distinguished three types of military commitments – so-called "circles of defense": (1) perimeter, (2) intra-frontier, and (3) remote military commitments. Perimeter defense denoted conventional military threats to Israel's territorial integrity vis-à-vis large standing Arab armies in the immediate vicinity of Israel's frontiers (i.e., Egypt, Syria, Jordan); intra-frontier commitments referred to defense within Israel's territory principally against terrorist attacks and low-intensity incursions; and remote military commitments stipulated contingencies and threats in a considerable distance from Israel such as Iraq and Iran (Cohen and Inbar 1991). During the Cold War, the predominant focus was on the "perimeter circle" that defined the frontlines of superpower rivalry in the Middle East and stipulated major conventional threats relevant to Israel's basic security. Yet, with the changes in the world order and systemic balance of power brought by the end of the Cold War in the early 1990s, Israel's strategic outlook has been shifting to a mixture of intra-frontier and remote military commitments. The Soviet Union, traditionally serving as a principal patron of Arab states, ceased to provide arms, training, and financial support to its regional proxies, leaving states such as Syria isolated with no resupplier, cash, or credit to modernize its large (and increasingly outdated) conventional forces. The end of the superpower rivalry coupled with the increasing fragmentation of the Arab world has essentially widened the qualitative gap between Israeli and Arab armies and, at least in the short term, reduced the prospects for high-intensity conventional war at the perimeter circle. According to Inbar, "the collapse of the Soviet empire is seen by the Israeli leadership as creating a new international atmosphere and is often credited for opening the door to Mideast peacemaking. Israel's adversaries lost their Soviet umbrella, a politico-military relationship that was an important factor in the Arab ability to confront Israel" (1996, 45). This meant that the potential of forming an Arab coalition that would simultaneously threaten or initiate a coordinated attack on Israel has diminished. At the same time, however, while the magnitude of conventional threats from the "inner ring" countries has relatively subsided in the 1990s, the development of regional asymmetrical threats – that is, long-range ballistic missiles, has enabled the "outer ring" countries, primarily Iran, to directly threaten Israel (Inbar 2002). In other words, Israel's conventional qualitative superiority has been offset or reduced by the increasing asymmetric capabilities and non-linear threats of the neighboring Arab countries. Indeed, writings in the official *IDF Journal* in the early 1990s warned of the impending "missile age" with the introduction and proliferation of new generation of Soviet-exported SSMs (i.e., SS-21) in the Arab arsenals and their strategic and operational implications for Israel (Levran 1990; Steinberg 1990)·

The transformation of Israel's external security environment during the 1990s further shifted Israel's threat perceptions, strategic outlook, and intensified Israel's security dilemmas. Increasingly, selected tenets of Israel's traditional security concept (i.e., rapid military decision) have been questioned along with the means to achieve them (Kober 1995). The perennial questions driving the debate focused on how to respond to the range of emerging complex challenges, how to allocate resources in order to retain existing qualitative military supremacy while deterring mounting asymmetric, low-intensity, and non-linear security threats.

It is within the context of these developments that the conceptual adoption and adaptation of the RMA gradually gained momentum in Israel's broader security debate. In particular, from the early 1990s, select Israeli defense analysts began to acknowledge the notion of RMA under the rubric of "future battlefield" (*sdeh hakrav haatidi*) (Adamsky 2010, 98). The Israelis analyzed U.S. strategic narratives, shaped by the experiences of the Gulf War, while observing the combat potential and implications of new military technologies (Brom 1999). Notwithstanding the initial Israeli skepticism toward the "Desert Storm" and its broader military significance, "Israelis have begun to recognize that there is indeed a revolution in progress" (Cohen, Bacevich, and Eisenstadt 1998, 92).

In this context, the pace, character, and direction of the Israeli RMA adaptation has been gradual. According to Uri Bar-Joseph, while "the IDF underwent a significant conceptual change after 'Desert Storm' and became more willing than before to structure its ground forces (in addition to its air and naval forces) around advanced systems" (1998a, 59). However, Israel's budgetary problems coupled with continuing conventional threats precluded "immediate radical shift in the buildup of the IDF's ground and air-to-ground capabilities" (Bar-Joseph 1998a, 59). In order to procure and exploit the full potential of RMA technologies, the IDF would need a comprehensive restructuring along the lines of a smaller and smarter (and possibly professional) force. This theme propelled another debate on updating Israel's military doctrine, organizational structure, and operational conduct (Cohen 1995). The origin of the "smaller, smarter force" is often credited to Lt. Gen. Dan Shomron (IDF CGS, 1987–91), who stated that "we are progressing in certain direction: we have to develop and acquire original 'smart' weapons systems that will allow us to maintain and improve our ability to achieve victory. As an inevitable consequence, in view of the high price of these weapons, the IDF will have to become smaller and more efficient" (Pedatzur 1991, 35).

Perhaps most importantly, a new and initially influential school of thought emerged in the mid-1990s within the Operational Theory Research Institute (OTRI). The OTRI, established together with the School for Operational Command in 1995 with the backing of IDF General Staff and the Central Command (CENTCOM), embarked on a process of organizational and conceptual transformation within the IDF that eventually translated into new concepts of operations, new methods of operational assessment, and, perhaps most importantly, a period of new strategic awareness and military thought. Its founders, Brig. Gen. (ret.) Shimon Naveh and Brig. Gens. (ret.) Dov Tamari and Zvi Lanir, believed in the need to critically rethink essentially all aspects of knowledge about war fighting and train future IDF commanders to think creatively, critically, systematically, and methodologically (Glick 2006). They argued that Israel's conservative conceptual innovation stemmed from the systemic absence of operational thinking coupled with doctrinal stagnation and inertia, preventing the IDF from adapting emerging technologies to new geopolitical realities and leading to frequent operational failures (Naveh 1996; Adamsky 2010, 99). Set up as a theoretical laboratory for training of senior military staff, OTRI's curriculum and methods departed sharply from established IDF frames of reference – its intellectual thrust focused on understanding the complexities of warfare in a nonlinear operational environment, in which the enemy, technology, and environment formed a complex adaptive system (Vego 2009). This, in their view, precluded the application of traditional decision-making processes, assessment methods, and operational planning. In other words, formulating a military doctrine for the constantly changing environment bound with complex uncertainties required new adaptive, systemic, learning strategies. Recognizing the limitations of traditional military terminologies, OTRI turned to alternative disciplines that, in their thought, offered greater relevance for assessing combat in nonlinear environments. Naveh and his colleagues were heavily influenced by the confluence of (1) postmodernist schools of thought, (2) architectural design, and (3) general systems theory as a methodology of inquiry and conceptualization (Weizman 2006a). By applying postmodern philosophy into operational military domain, OTRI attempted to converge the general systems theory with operational art – the design and execution of operations that formed an intermediate field between strategy and tactics, which they argued was missing in the Israeli military experience. To bridge this gap, OTRI studied and emulated Soviet operational concepts and doctrines (i.e., Soviet theory of deep operations), which they viewed as intellectually and theoretically superior to the American ALB concepts and applied them to Israeli military affairs and doctrine (Naveh 1997). At the same time, they studied, translated, and disseminated articles on the U.S. RMA theories and organized meetings and seminars with their American counterparts at the ONA, including Andrew Marshall. According to Adamsky, "Israelis who participated in the meeting saw themselves as the architects of the Israeli version of the RMA … the subsequent contacts, workshops, and scientific seminars on military theory, innovation, and experimentation with the ONA became one of the most important sources of intellectual influence on the builders of the Israeli RMA concept" (2010, 103). At the same time, OTRI shaped strategic discourse in the United States, particularly by disseminating its research on new assessment methodologies and understanding the operational level of war. According to Lt. Col. David Pere, co-author of the Marine Corps' operational doctrine, "OTRI's influence on the intellectual discourse and understanding of the operational level of war in the U.S. has been immense. The U.S. Marine Corps has commissioned a study of design that will result in a Marine Corps Concept of Design that is based heavily on Naveh's work" (Glick 2006).

The net result of OTRI's research and educational activities was two-fold: (1) a theory of SOD that later served as a conceptual and doctrinal basis for the formulation of Israeli RMA-oriented operational concept titled "IDF Concepts of Operations"; and (2) cognitive tools and methods of operational assessment that were adopted by the regional commands and the General Staff (Glick 2006; Adamsky 2010). The SOD methodology, in particular, aimed to create a "structured thought process" that would facilitate a systemic disruption of the enemy through the use of operational shock, continuous learning, and adaptation. Rather than focusing on traditional force-on-force destruction achieved by a series of actions, the SOD searched for ways to transform the relationships and interactions between the entities within a system. In his book, Adamsky explains that the SOD envisioned the enemy as a multidimensional, adaptive, and open system whose ability to function would be neutralized by (1) *fragmentation strike* – isolating an enemy's military subsystem from the strategic super-system and disrupting its logic; (2) *simultaneous coordinated actions* across the spectrum of operations – in order to shock and paralyze an enemy system; and (3) *sustained momentum* – that would exploit the synergistic effects resulting from fragmentation and simultaneity and, in doing so, deny the opposing system response time and thereby accelerate its eventual collapse (2010, 101). Inherently, the SOD concept challenged IDF's traditional operational planning methods. Instead of emphasizing on regressive, solution-oriented action, uniform templates, formal patterns, and problem solving, the SOD focused on a holistic (systemic) and flexible planning method, emphasizing continuous learning, problem setting, and reframing of the relationships and various parts of the enemy system. Under the SOD, IDF commanders would become "operational architects" (Weizman 2006a) capable of sensing, learning, interacting with their environment, and, ultimately, adapting to complex missions (Czege 2009). Recognizing that the enemy's system will eventually change and adapt, no strategic directive under the SOD would be final or complete. Instead, commanders would conceptualize "knowledge maps" based on an interactive "discourse" or "critical dialogue" between political echelons and operational and tactical command levels, which would build understanding through a recursive dialectical process and fuse the various situational insights(Adamsky 2010, 102).

During the late 1990s, OTRI's SOD-oriented research, methodologies, and education gradually permeated in various levels into senior ranks of regional commands (i.e., Brig. Gen. Moshe Ya'alon in the CENTCOM at that time strongly supported the concept), then to the General Staff, and subsequently throughout the IDF. At the same time, however, while the SOD became one of the major sources for a new planning process, it also generated "huge" debates within the IDF, both formally and informally. From the perspective of its supporters, the SOD methodology was there not only to help IDF commanders to plan their campaigns but also, perhaps more importantly, to propel transformation in the IDF's learning methods and intellectual culture. For its opponents, these ideas soon proved to be an illusion (Kober 2011).

2000s: subconventional warfare and SOD debate

At the turn of the millennium, the IDF was increasingly engaged in LICs – military operations in urban areas to combat terrorism – to detect and stop attacks on Israeli population, thereby reducing its scope, depth, and damage. During the five years of "Second Intifada" (2000–2005) in the West Bank and the Gaza Strip, the IDF focused on managing primarily asymmetric and LIC challenges. The Israelis defined these in the context of "sub-conventional warfare" – a multifaceted term embracing LICs, counterterror operations, and various forms of guerilla warfare (Eshel 2006). In 2004, then the IDF CGS Lt. Gen. Moshe Ya'alon noted that Israel has been gradually moving into crossfire of two extreme threats of a threat scale: (1) a *sub-conventional LIC* against terrorism, on one hand, and (2) *supra-conventional threat* emanating from long-range ballistic missiles and potential regional proliferation of WMD, on the other (Ben-David 2004b, 19). With shifting strategic and operational priorities, the IDF had to examine changes necessary for developing new operational concepts in relation to new threats, weapons and technological developments, low- as well as high-intensity conflicts, asymmetric warfare, and trends in other advanced militaries (O'Sullivan 2004). While the learning process started much earlier during the late 1990s (i.e., CGS Shaul Mofaz 1998–2002), it accelerated in substance under the CGS Lt. Gen. Ya'alon (2002–5). Under Ya'alon, the IDF organized evaluation teams, meetings of generals and heads of regional commands, and working groups in all branches and levels of the IDF specifically with the aim to reassess various aspects of the Israeli military doctrine and operational concepts (Ben-David 2004a, 27). In the process lasting over two years, the IDF relied on OTRI's professional and academic support particularly with teaching SOD methodology and coursework, as well as input and lessons learned from combat units (Adamsky 2010, 105). OTRI's staff argued that IDF needs an adaptable doctrine or concept for fighting complex missions in urban areas. According to Brig. Gen. Aviv Kohavi, "when the fighting [Second Intifada] started we realized there were no relevant doctrine or techniques for fighting LIC combat in populated urban areas ... With the situation in constant flux, on a daily basis, were required to develop solutions to unique situations" (Ben-David 2004a, 27). It is in this context that OTRI's SOD became increasingly influential in the IDF's operational conduct, particularly in LIC's mission cycles in urban areas. For example, in 2002 during operation "Defensive Shield," Gen. Kohavi, in charge of operations in the city of Nablus, applied one of the SOD concepts of "inverse geometry," which he described as the "reorganization of the urban syntax by means of a series of micro-tactical actions" (Weizman 2006b). Drawing heavily on the SOD methodology, the concept attempted to reinterpret the linear views of urban space where movement is limited to predetermined paths (i.e., streets, roads, alleys, courtyards, doors) into a "flexible medium of warfare." The IDF would "maneuver through walls" across the entire depth of the city and from all directions in order to surprise the enemy by avoiding the heavily fortified streets, mined alleys, road blocks, remote-controlled explosive charges, booby-trapped doors and windows, sniper fire, and also to prevent casualties (Ben-David 2004a, 27). Organized in small units, IDF soldiers moved horizontally through walls and vertically through blasted ceilings and floors, swarming the city from three directions simultaneously and cutting across a dense and contiguous urban structure through hundreds of meters of "overground tunnels." This method created a considerable element of surprise, shock, and disorientation of the enemy combatants, who had expected the IDF to launch air strikes before entering the city with infantry (Weizman 2006b).

Through the continuous LIC engagements, the IDF was able to improvise, adapt, and ultimately innovate its counterterror tactics through the networking of its C4ISR systems, precision firepower capabilities, and Special Forces (Hughes 2004, 34). The IDF aimed to depart from previous reactive responses to terrorism toward *preemptive* active strategies and tactics against terrorist infrastructures and leaders (i.e., targeted assassinations) (Eshel 2002a, 20). The buzzwords within the IDF became "intelligence-driven operations," "integrated-situational assessments," "sensor-to-shooter cycles." Joint intelligence operation centers, created in the late 1990s, were transformed into sophisticated C4ISR centers that combined representatives from all relevant IDF branches and units, which provided a continuous flow of information to all territorial commands (Ben-David 2004a, 27). By integrating relevant information from real-time intelligence operations, continuous analysis of the threat, and aerial precision attack capabilities, the IDF was able to reduce the decision "sensor-to-shooter" cycles to a few minutes between the time a target appears, is positively identified, and destroyed (Eshel 2006). By 2004, then the Defense Minister Shaul Mofaz acknowledged that the IDF is successful in thwarting more than 90 percent of planned attacks against Israel (Eshel 2002b, 24).

In essence, the operational experience and relative success of the LIC missions from nearly five years of continuous antiterror urban warfare coupled with OTRI's SOD-oriented methodology has gradually transformed the IDF doctrine, operational concepts, weapon system development, and overall defense management. In 2004, and after several alterations and revisions, Brig. Gen. Moshe Ya'alon presented an internal draft of a new field manual for the army, along with the multi-year KELA 2008 force modernization plan (Ben-David 2004d). However, the relative complexity in language, tactics, methods, and procedures generated heated debates on its utility and applicability in practice. In particular, as Adamsky explains, "the new concepts generated different meanings for different command authorities within the IDF with different conceptual backgrounds. However, many senior IDF officers subscribed to the new concepts purely from the bureaucratic and intellectual conformism of the moment; without actually understanding the concepts. This has led to a state of 'conceptual disorientation' within the IDF" (2010, 109).

Then in April 2006, a newly appointed CGS Lt. Gen. Dan Halutz inherited the new combat doctrine and signed it as a *Concept of Operations* (CONOP). The 2006 CONOP departed sharply from previous operational doctrines – it combined selected elements of U.S. defense transformation, that is, NCW, effects-based operations (EBOs), diffused warfare (DW), as well as OTRI's SOD methodology. It was designed to cover strategy, force transformation as well as introduce a new military language, new structure for staff methodology, battlefield analysis, and the structure and content of orders (Vego 2009, 72). In other words, the 2006 CONOP provided an alternative "conceptual framework" for conducting operations in the context of *both* nonlinear LICs and high-intensity conventional warfare. It represented the culmination of Israeli RMA-oriented military thought – it was the first time the IDF presented a codified written form of a field manual for the operational and strategic levels of war, reflecting "the quintessence of Israeli views on the nature warfare in the RMA era" (Adamsky 2010, 109). CONOP relied heavily on the SOD methodology, particularly in attempting to identify and attack "the rationale of the opponent system" by creating an "effects-based" campaign consisting of a series of "physical and cognitive appearances" designed to influence the consciousness of the enemy rather than destroying it – that is, creating conditions of "cognitive perception of defeat" (Ben-David 2007; Kober 2008). Indeed, the 2006 CONOP assumed that a physical defeat of the enemy is impossible, or even unnecessary, in the current types of LICs. Instead of capturing the enemy territory and achieving rapid "battlefield decision" by linear means, the IDF should create conditions to "contain and disrupt the enemy system" through a "protracted fatigue," "shock paralysis," and "superior determination." While the idea of maneuver warfare was not completely discarded, the IDF would limit the grand maneuver by armored formations and place the emphasis on "aerial dominance." Specifically, the IDF would transfer the war into enemy territory by applying "stand-off decisive precision firepower" supported by advanced C4ISR platforms and real-time intelligence. The reliance on technological capabilities would minimize casualties, shorten the sensor-to-shooter cycle, and attain operational objectives in the shortest time possible. Such capability would lead to a concept of "empty battlefield" – destruction of targets without a need to enter the battle areas. Influenced by U.S.-centric EBO concepts, the CONOP also envisioned "swarmed attacks" – coordinated attacks by separate but synchronized/networked units (i.e., light infantry, Special Forces, air force, and intelligence units) operating throughout the battlefield. A traditional "unit of action," the basic force structure building block, would be replaced by networked "dynamic molecular forces" – semiautonomous units with advanced sensor and shooter components capable of conducting various missions that would complement other molecular systems in its proximity (Yaari and Assa 2005). The "molecule's" lack of massive signature coupled with its dynamic and versatile attributes would make it difficult for the enemy to detect and predict its patterns of conduct. As Adamsky noted, "a network of separate units would be diffused through the whole operational depth and would operate semi-autonomously, but jointly in frames of unified campaign plan. Under the assumption that it takes a network to combat a network, these 'swarming' formations would adjust themselves to the stealth capability of the enemy" (2010, 106).

Three months after its formal adoption, Israel launched a 33-day military campaign in Lebanon, responding to the kidnapping of two IDF servicemen by Hezbollah. Notwithstanding the broad international support, military superiority, and tactical achievements on the ground in the early stages of the campaign, the IDF was unable to neutralize Hezbollah's concealed bunker system, disrupt its decentralized command and control system, and, most importantly, prevent short-range rockets to be launched on Israel. The Second Lebanon War was widely perceived in Israel as one that failed to achieve its objectives. Among the many critical explanations for the shortcomings in the combat performance of the IDF during the Second Lebanon War, as outlined in the official Winograd Committee Report, was the doctrinal confusion, tensions, and operational distortion generated by the new CONOP. Its critics described the CONOP as "completely wrong … countering the basic principles of operating an army in general and the IDF in particular" (Ben-David 2007). Some senior IDF officers pointed to the absence of clear professional language and uniform command procedures. The 2006 CONOP replaced traditional principles, notions, and frames of reference such as "objective," "initiative," "persistence," "concentration of effort," or "center of gravity" with terms such as "campaign rationale," "controlling territory," and "conscious-burning" of the enemy. As such, there were problems with the way the commands were issued, which did not provide the coherent distinction between objective, mission, and method. Others argued that "Israeli military doctrine and planning have tended to adopt false assumptions and beliefs ... particularly with American RMA-oriented technology-based approach" (Kober 2008, 17). In sum, there were many political, institutional, organizational, and cultural impediments for the SOD. According to Matt Matthews, "not every officer in the IDF had the time or the inclination to study postmodern French philosophy ... Many IDF officers thought the entire program elitist, while others could not understand why the old system of simple orders and terminology was being replaced by a design that few could understand" (2006, 25).

However, Israeli RMA proponents point out that the SOD-oriented 2006 CONOP has not been really implemented in the Second Lebanon War – its introduction just prior to the war meant that it has not fully permeated into the IDF's training, organizational structure, and operational conduct, and thus it could not be implemented. Shimon Naveh believes that CONOP has been widely misunderstood: "I don't think that Lebanon exemplified [CONOP's] failure ... if you read the testimonies of the prime minister and the defense minister to the Winograd Committee, you see that they understood a few things, but did not understand them in depth" (Feldman 2007). Regardless of the debate, with the limited Israeli achievements in Lebanon, the IDF initiated comprehensive review of the CONOP doctrine while addressing organizational, structural, and operational deficiencies. Indeed, following the experiences and lessons learned in the operations "Cast Lead" in Gaza (2009), "Pillar of Defense" (2012), and "Protective Edge" (2014), the IDF began implementing new "Firepower and Combined Arms Concepts" – projected to form a new baseline for future IDF strategy, force structure, and operational conduct (Opall-Rome 2010). Based on available reports, the new concepts depart sharply from the traditional notions and rules governing high- or low-intensity conflict, air power, ground- and sea-based strikes. The IDF does not distinguish between state or nonstate actors and does not acknowledge the notion of "hybrid" or "mosaic" wars. Instead, Israeli defense planners view firepower regardless of its source, range, or launching platform. The implementation of firepower is *target–driven*, network-enabled, and precision-oriented. For example, IDF ground commanders can assume instant "organic control" of any available "shooter" assets of the air force as well as the navy's ship-launched missiles and artillery as soon as targets are identified. Inherently, the new concepts reflect Israel's continuous drive to search for operational solutions relevant to the progressive complexity of emerging security challenges.

2010s: cyber-kinetic operations

Within the context of evolving asymmetric challenges, the debates on conceptual adoption and adaptation of the cyber-oriented security gradually permeated into Israel's broader strategic debates. In the early 1990s, during the IDF CGS Ehud Barak's tenure, select Israeli defense analysts began to acknowledge the notion of "cyber activity" under the rubric of "future battlefield" (*sdeh hakrav haatidi*) (Adamsky 2010, 98; Ben-Israel 2012). They analyzed the emerging trajectory of information technologies used in combat while observing the potential and implications of new generation of PGMs, cruise missiles, command-and-control systems, integrated intelligence, and electronic warfare (Brom 1999, 13–18). At that time, cyber security in the military domain was conceptualized along the lines of "information warfare" as a sphere of decisive importance in which achieving superiority in relation to the rival was seen as the key to deciding military conflicts. Over the past decade, however, Israel's cyber security issues, aspects, and policies have shifted with the increasing realization of two key assumptions: (1) the accelerating expansion of cyberspace has increased political, military, and socioeconomic dependencies on the cyber domain by an order of magnitude; and select adversaries could in theory disrupt, destroy, or subvert key strategic targets (i.e., critical infrastructures) without confronting the defending armies and without exposure and clear attribution. (2) Existing civil/military compartmentalized organizational structures, responsibilities, policies, and regulations for protecting computerized systems are not adequate to enable a comprehensive defense vis-à-vis the continuously evolving challenges and threats in cyberspace. In this context, "cyber" debate within and outside the Israeli defense establishment has shifted on the emerging threats, challenges, as well as opportunities of cyberspace as a new medium for *civil–military strategic interactions.*

In particular, as more critical information infrastructure systems require telecommunications, clouds, and computers connected to the Internet or proprietary networks, there is a growing awareness in Israel that different types of adversaries may seek to influence strategic outcomes by accessing and altering both the systems themselves and the data that reside within. In other words, an increased dependence on cyberspace by individuals, companies, and organizations amplifies the vulnerabilities and potential for harm – not only of information but also to persons, property, and functional continuity of the state. Moreover, the realities of Israel's security environment predicate evolving cyber conflicts occurring not only during wartime or crises but on a persistent basis – confrontations in and out of cyberspace, including cyberattacks on computerized systems, physical systems, and processes controlling critical information infrastructure, information operations, and various forms of cyber espionage (Even and Siman-Tov 2012). In October 2013, for example, IDF CGS Benny Gantz described what he believes will characterize Israel's future wars:

> Along with the border battles, which will also have serious implications for the Israeli civilian rear, "a vast cybernetic war will rage that will affect not only the military but also the civilian systems." It will be an "almost transparent" war, as media on both sides will cover it intensively in real time.
>
> (Harel 2013)

During the recent IDF operation code-named "Protective Edge" in Gaza in 2014, Israel faced large-scale cyberattacks on its civilian communications infrastructure, including distributed denial of service attacks and domain network system attacks from both state and nonstate actors, traced to Qatar and Iran – Hamas' main benefactors. Cyber attackers also targeted the IDF's websites and communications networks. The Israeli Security Agency (*Shin Bet*) announced that these attacks against government and military networks had been contained, while in the civilian sector the attacker's intent to cause maximum disruption was not achieved. Israel's strategy in responding to such threats is not so much about select advanced cyber defense systems and technologies but developing unique interdisciplinary methodologies in a multidisciplinary national "cyber ecosystem" that integrates national labs, military intelligence units, C4I organizations, National Cyber Command, and start-up firms and entrepreneurs. In doing so, Israel is developing "a national cyber defensive envelope" – a multilayered cyber defense strategy leveraging automated computerized systems and highly trained personnel that proactively combine intelligence, early warning, passive and active defence, and offensive capabilities across civil–military domains.

At the core of Israel's cyber capabilities and sustained innovation drive, whether in the civil, military, or commercial sectors, is the selection, training, research and development, skills and service experience of "cyber defenders" in the IDF. Indeed, the IDF is often credited in the creation and sustained success of the Israeli high-tech industry – by creating not only the key mechanisms (i.e., high skill training and its spin-off effects) but serving as "the main node in the national innovation system that diffuses information, spurs collective learning, and creates standards for the entire industry" (Breznitz 2002, 6). The process begins with the IDF identifying and recruiting suitable candidates excelling in cybertech subjects at select top high schools. Based on a range of qualifying exams, these students are then placed either at (1) the Mamram (*Merkaz Mahshevim UMa'arahot Meida*) – the Center of Computing and Information Systems, which is the IDF's central computing system unit, providing data processing services for all arms and the General Staff of the IDF, and its related unit, or (2) the School for Computer Professions (*Basmach*). Following graduation, Basmach students go on to serve in various IDF Military Intelligence and Manpower Directorate units, while some graduates are often offered a position in Mamram itself.

While specific details of IDF's cyber units and capabilities are difficult to ascertain from open sources, well-known units of the IDF that specialize in various aspects of cyber defense (and offense) are frequently profiled in the media for their high levels of operational sophistication and their cutting-edge training of personnel. Among the most publicized are Intelligence Corps Unit 8200, which deals with SIGINT and code decryption (Ingersoll 2013); the Cyber Unit within 8200, established in 2009; the C4I Directorate, leading network-centric warfare (Dombe 2014a); and its sub-unit, the Cyber Defense Division, responsible for preventing and detecting infiltrations into military networks (Katz 2012). In 2013, the IDF consolidated all aspects of its cyber situational awareness and command activities into a new cyber HQ, which is linked with the civilian Tehila (governmental internet infrastructure) system, the E-Government project, and the recently established Cyber Security Authority. In the same year, the Israeli ministry of defense announced the establishment of a new cyber directorate at Mafat (the MOD Directorate of Defense R&D), which is tasked to conduct cyber R&D activities for select branches of the Israeli defense establishment and, perhaps more importantly, integrate cyber innovation and capabilities of Israel's commercial high-tech sector into defense (Heller 2013).

Operational aspects of these IDF cyber units are clouded in secrecy. However, press reports have linked Israel with the development of the *Stuxnet* virus, which disabled nuclear centrifuges in Natanz in 2010 (Sanger 2013). An additional example of Israel's operational cyber capabilities is "Operation Orchard," an Israeli airstrike on a Syrian nuclear facility in the Deir ez-Zor region during the night of September 6, 2007 (Hersh 2008). In this operation, the IDF combined elements of classical air power with innovative cyberattacks that paralyzed Syria's air defenses. "Operation Orchard" is a prime example of the type of capabilities that the IDF may leverage in the future in utilizing cyberspace as a force multiplier:

- detection of future threats through sophisticated, cyber-enabled intelligence gathering, including satellite monitoring;
- real-time intervention in enemy weapons systems or defensive systems, such as the Syrian air defense system in this case; and
- utilization of traditional air, sea, or ground power in conjunction with cyber capabilities.

The deterrent effect of such mixed "kinetic-cyber" operations is not yet well understood, but one report of "Operation Orchard" noted that it restored Israel's credibility as a deterrent against Syrian forces and also served as an effective signal to Iran. A more recent example of Israel's use of its defensive cyber capabilities is the capture of the Iranian arms ship Klos-C in the Red Sea in March 2014 in Operation "Full Disclosure." The successful naval interception was carried out 1,500 km from Israeli shores and was enabled as a result of the "advanced cyber and communications capabilities" (Dombe 2014b).

The progressive complexity in the cross-domain interactions between kinetic and cyber threats coupled with constraints imposed by existing cyber defense organizational structures has shifted the debate in the Israeli cyber security policy community. In 2002, Israel passed the National Security Ministerial Committee Resolution 84/B regarding the responsibility for protecting civilian computerized systems in the state of Israel. The Resolution 84/B became de facto the national civilian cyber defense policy, providing the initial framework for national Critical Computer Systems policy (Tabansky 2013). At that time, Israel defined 19 "critical" systems, in both public and private domains, and dictated a "shared responsibility" for protecting their computerized systems between its users and regulators. In practice, a "user" – an organization – would be tasked with financing, protection, maintenance, upgrading, backup, and recovery of its critical IT systems while sharing the information with the "regulator" – that is, existing chiefs of security at government ministries. An oversight body, under *Shabak*, was created in the form of National Information Security Authority (NISA; *Re'em*). After nearly a decade, however, the baseline perspective and framework of the Resolution 84/B – the responsibilities, authorities, and functions of the governmental division and "special bodies" responsible for protecting critical civilian computerized systems – has reached significant political and legal constraints. In particular, civilian systems and networks that have not been defined as essential became highly vulnerable to cyber threats, while the organizational responsibilities for the protection of computerized systems were deemed compartmentalized and fragmented (Levi 2011). Against this background, in November 2010, Prime Minister Benjamin Netanyahu appointed the Chairman of the National Council for Research and Development, Prof. Gen. (ret.) Isaac Ben-Israel, to review existing policies and formulate a national plan for dealing with the growing cyber threat – the National Cyber Initiative. Netanyahu also emphasized the need to position Israel in the leading five nations in the cyber field by 2015, in accordance with a vision of Israel that maintains its global position as a center of information technological development with powerful cyberspace capabilities. The detailed report was eventually published in May 2011, summarizing the work of 80 experts. Eight sub-committees consisting of senior decision makers and representatives – including the MOD Directorate of Defense R&D, the chief scientist of the ministry of economics, the National Economic Council in the prime minister's office, the ministry of finance, the ministry of science and technology, the Unit 8200 and the Unit for Telecommunications and Information Technology in the IDF, the Counter-Terrorism Headquarters in the National Security Council, NISA, the Atomic Energy Commission, and other experts from the military, academia and government ministries – convened over six months for discussions to draft recommendations to the prime minister and cabinet. A task force condensed these recommendations into a draft Government Decision 3611 to establish the National Cyber Bureau (NCB), which was passed unanimously on August 7, 2011. The multidisciplinary nature of the initiative, and especially the tripartite cooperation among senior military, academia, and government personnel, set the tone for the second phase of national engagement with cyber defense policy. Moreover, the end result was an operational bureau within the prime minister's office, which began its work formally in early 2012. Paradoxially, for the next two years, the NCB has been opposed and debated by the Shin Bet, internal security agency, which argued that action against hackers should be taken proactively in the early organization and planning stages, rather than reactively. The Shin Bet claimed that the NCB is unable to carry out this task because it lacks intelligence-gathering capabilities, has no operational tradition of deterrence, and no possibility of integration with similar security organizations worldwide (Barak 2014). After nearly two years of policy battles, Prime Minister Netanyahu announced on September 21, 2014, the establishment of a new government authority alongside the NCB with the responsibility for protecting Israel's economy and civilian space from computer attacks, effectively rejecting recommendations of the Shin Bet security services.

At the time of this writing, three of the lead stakeholders in Israel's national strategy to develop cutting-edge cyber and dual-use R&D have been the MOD's R&D directorate (MAFAT), the chief scientist at the ministry of the economy, and the ministry of defense's Defense Export Controls Agency, which is responsible for implementing the Defense Export Control Law of 2007. The successful implementation of several policy initiatives that these agencies and others have promoted within Israel's national cyber strategy has leveraged the multidisciplinary approach reviewed above. As a result, Israel's cyber and dual-use exports has in mid-2014 reached US$ 3 billion, second only to the United States, and constituting 5 percent of the global market, according to the NCB. Moreover, "Israeli firms last year [in 2013] raised US$165 million in investment funding, a figure [which] represents 11 percent of global capital invested in the field of cyber. According to NCB data, 14.5 percent of all the firms worldwide attracting cyber-related investment are Israeli-owned" (Opall-Rome 2014). In this context, Israel is presently in the third stage of framing a national cyber defense vision. The missing element until now has been the private sector, which had only participated on the fringes of public defense policy discussions in the two earlier stages. The present national effort in the R&D and trade actively promotes the inclusion of this sector by adding it proactively to the cooperative efforts of the existing interdisciplinary strategy. A recent initiative within the prime minister's office, called Digital Israel, will further expand government–private sector collaboration as an integral part of the national vision. Furthermore, as part of a major national initiative to develop Israel's south, represented by the IDF's massive relocation to the region over the next few years, Israel is developing the Beersheva Municipality as a leading cyber R&D hub. The Beersheva facility includes leading cyber industries such as EMC, Lockheed Martin, Deutsche Telekom, IBM, and JVP; cutting-edge industrial academic research in the field of information security; leading government agencies such as the Cyber Bureau and the national CERT program; and next-generation educational frameworks.

In conclusion, Israel's military innovation over the past two decades has been driven by a constant search for effective responses to a constantly changing security environment. A prevailing characteristic has been the need for a proactive, multidisciplinary, and intersectorial commitment in both civil–military domains aimed to respond to particular threats. While Israel continues to refine its strategies, technologies, organizations, and operational concepts – some of which may be ahead of military-technological superpowers, it is appropriate to insert a word of caution about the Achilles heel of all innovative military and civilian capabilities. The price of success of innovative systems is the perennial effort to continuously innovate. The various stakeholders in Israel's security must, therefore, fully commit and sustain this effort. Without a continuously updated, well-formulated national strategic policy and resource allocation, coupled with training and operational experience of its armed forces, ensuring effective responses to the next generation of threats may be at risk.

References

Abramowitz, Jeff. 1988. "The Evolution of the Ground Corps Command." *IDF Journal* 3 (3): 8–14.

Adamsky, Dima. 2010. *The Culture of Military Innovation: The Impact of Cultural Factors on the Revolution in Military Affairs in Russia, the US, and Israel.* Palo Alto, CA: Stanford University Press.

Allon, Yigal. 1969. *Curtain of Sand (Hebrew).* Tel Aviv, Israel: Ha-Kibbutz Ha Mevchad Press.

Barak, Ravid. 2014. "Battle Move in Israel's Cyber Turf War: Shin Bet Loses Authority Over Civilian Space." *Haaretz*, September 21.

Bar-Joseph, Uri. 1998a. "Israel's Northern Eyes and Shield: The Strategic Value of the Golan Heights Revisited." *Journal of Strategic Studies* 21 (3): 46–66.

——. 1998b. "Variations on a Theme: The Conceptualization of Deterrence in Israeli Strategic Thinking." *Security Studies* 7 (3): 145–81.

——. 2000. "Towards a Paradigm Shift in Israel's National Security Conceptions." *Israel Affairs* 6 (3): 99–114.

——. 2004. "The Paradox of Israeli Power." *Survival* 46 (4): 137–56.

——. 2007. "Israel's Military Intelligence Performance in the Second Lebanon War." *International Journal of Intelligence and Counterintelligence* 20 (4): 583–601.

——. 2008. "Lessons Not Learned: Israel in the Post Yom Kippur War Era." *Israel Affairs* 14 (1): 70–83.

Ben-David, Alon. 2004a. "A New Perspective: Israel Defense Force Rethinks Combat Doctrine." *Jane's Defense Weekly* 41 (35): 26–28.

——. 2004b. "Double Jeopardy: Israel is Caught in the Crossfire of Two Extreme Threats to Its Security." *Jane's Defense Weekly* 41 (46): 19–28.

——. 2004c. "Inner Conflict: Israel's Low-Intensity Conflict Doctrine." *Jane's Defense Weekly* 41 (35): 27.

——. 2004d. "Israel Adjusts Acquisition Plan." *Jane's Defense Weekly* 41 (26): 6.

——. 2007. "Debriefing Teams Brand IDF Doctrine 'Completely Wrong'." *Jane's Defense Weekly* 44 (1): 1–7.

Ben-Dor, Charles. 1989. "Combined Forces at Every Level." *IDF Journal* 18 (Summer Issue): 6–9.

Ben-Gurion, David. 1953. *Uniqueness and Destiny.* Tel-Aviv, Israel: IDF Publishing.

Ben-Israel, Isaac. 2012. *Introductory Remarks of the Annual Cyber Security International Conference.* Tel Aviv, Israel: Tel Aviv University.

Ben-Israel, Yitzhak. 2009. "National Security vs. Energy Indepdence in Israel." Lecture at Tel-Aviv University, May 21..

Beres, Louis Rene. 1995. "Israel's Bomb in the Basement: A Revisiting of Deliberate Ambiguity vs. Disclosure." *Israel Affairs* 2 (1): 112–36.

Bolia, Robert. 2004. "Overreliance on Technology in Warfare: The Yom Kippur War as a Case Study." *Parameters* 34 (2): 46–56.

Breznitz, Dan. 2002. *The Military as a Public Space: The Role of the IDF in the Israeli Software Innovation System.* MIT Working Paper IPC-02-004.

Brom, Shlomo. 1999. "Operation Desert Fox: Results and Ramifications." *INSS Strategic Assessment* 2 (1): 13–18.

Clary, David. 1988. *The Bekaa Valley: A Case Study.* Montgomery, AL: Air Command and Staff College.

Cohen, Avner. 1998. *Israel and the Bomb.* New York, NY: Columbia University Press.

Cohen, Eliot, Andrew Bacevich, and Michael Eisenstadt. 1998. *Knives, Tanks, and Missiles: Israel's Security Revolution.* Washington, DC: Washington Institute for Near East Policy.

Cohen, Stuart. 1995. "The Peace Process and Its Impact on the Development of a Slimmer and Smarter Israel Defense Force." *Israel Affairs* 1 (4): 1–21.

Cohen, Stuart, and Efraim Inbar. 1991. "A Taxonomy of Israel's Use of Military Force." *Comparative Strategy* 10 (2): 121–38.

Cordesman, Anthony. 2002. *Peace and War: The Arab–Israeli Military Balance Enters the 21st Century.* Wesport, CT: Praeger.

Creveld, Martin. 1998. *The Sword and the Olive: A Critical History of the Israeli Defense Forces.* New York, NY: Public Affairs.

Czege, Huba Wass. 2009. "Systemic Operational Design: Learning and Adapting in Complex Missions." *Military Review* 89 (1): 2–12.

Dayan, Uzi. 1999. "Air Power: The Israeli Perspective." *Military Technology* (5): 4–5.

Dombe, Ami Rojkes. 2014a. "Inter-Arm Tactical Communication." *Israel Defense,* June 29.

——. 2014b. "The IDF is Ready for the Cloud Challenge." *Israel Defense,* April 14.

Eshel, David. 2002a. "Israel Refines Its Pre-Emptive Approach to Counterterrorism." *Jane's Intelligence Review* 14 (9): 20–22.

——. 2002b. "Israel Hones Intelligence Operations to Counter Intifada." *Jane's Intelligence Review* 14 (10): 24–26.

——. 2006. "Israel Air Force Transforms for War Against Terror." *Military Technology* 30 (3): 21–29.

Even, Shmuel, and David Siman-Tov. 2012. *Cyber Warfare: Concepts and Strategic Trends.* Tel Aviv, Israel: Institute for National Security Studies.

Feldman, Shai. 1997. "Israel's National Security: Perceptions and Policy." In *Bridging the Gap: A Future Security Architecture for the Middle East,* edited by Shai Feldman and Abdullah Toukan, 7–32. Lanham, MD: Rowman & Littlefield Publishers.

Feldman, Yotam. 2007. "Dr. Naveh, or, How I Learned to Stop Worrying and Walk Through Walls." *Haaretz,* November 1.

Glenn, Russell. 2008. *Questioning a Deity: A Contemplation of Maneuver Motivated by the 2008 Israeli Armor Corps Association "Land Maneuver in the 21st Century" Conference.* Latrun, Israel: Israeli Armor Corps Association.

Glick, Caroline. 2006. "Column One: Halutz' Stalinist Moment." *The Jerusalem Post,* June 8.

Goodman, Hirsh, and Seth Carus. 1990. *The Future Battlefield and the Arab–Israeli Conflict.* London, UK: Transaction Publishers.

Grant, Rebecca. 2002. "The Bekaa Valley War." *Air Force Magazine* 85 (6): 58–62.

Handel, Michael. 1973. *Israel's Political-Military Doctrine.* Cambridge, MA: Harvard University Press.

——. 1996. "The Evolution of Israeli Strategy: The Psychology of Insecurity and the Quest form Absolute Security." In *The Making of Strategy: Rulers, States, and War,* edited by Murray Williamson, Alvin Bernstein, and MacGregor Knox, 534–79. Cambridge, UK: Cambridge University Press.

Harel, Amos. 2013. "Israel's Next War." *Haaretz,* October 12.

Heller, Mark. 2000. *Continuity and Change in Israeli Security Policy.* Adelphi Paper 335. 1–82. London, UK: The International Institute of Strategic Studies, 1–82.

Heller, Or. 2013. "New Cyber Directorate in Mafat." *Israel Defense,* December 3.

Hersh, Seymour. 2008. "Why Did Israel Bomb Syria?" *The New Yorker,* February 11.

Horowitz, Dan. 1987. "Strategic Limitations of a 'Nation in Arms'." *Armed Forces & Society* 13 (2): 277–94.

——. 1993. "The Israeli Concept of National Security." In *National Security and Democracy in Israel*, edited by Avner Yaniv, 1–11. Boulder, CO: Lynne Rienner Publishers.

Hughes, Robin. 2004. "Interview: Lt. Gen. Moshe Ya'Alon Israel Defense Force Chief of Staff." *Jane's Defense Weekly*, November 17, 34.

IDF Doctrine and Training Division. *IDF Dictionary* [Hebrew]. Publication 1-10.1998.

Inbar, Efraim. 1983. "Israeli Strategic Thinking After 1973." *Journal of Strategic Studies* 6 (1): 36–59.

——. 1996. "Contours of Israeli New Strategic Thinking." *Political Science Quarterly* 111 (1): 41–64.

——. 1998. "Israeli National Security 1973–96." *The Annals of the American Academy of Political and Social Science* (558): 62–81.

——. 2002. "Israel's Strategic Environment in the 1990s." *Journal of Strategic Studies* 25 (1): 21–38.

Inbar, Efraim, and Shmuel Sandler. 1994. "Israel's Deterrence Strategy Revisited." *Security Studies* 3 (2): 330–58.

Ingersoll, Geoffrey. 2013. "The Best Tech School on Earth is Israeli Army Unit 8200." *Business Insider*, August 13.

Karpin, Michael. 2006. *The Bomb in the Basement: How Israel went Nuclear and What That Means for the World.* New York, NY: Simon & Schuster Publishers.

Katz, Yaakov. 2012. "First IDF Cyber Defenders Graduate." *Jane's Defense Weekly*, April 20.

Kober, Avi. 1995. "A Paradigm in Crisis? Israel's Doctrine of Military Decision." *Israel Affairs* 2 (1): 188–211.

——. 2001. "Has Battlefield Decision Become Obsolete? The Commitment to the Achievement of Battlefield Decision Revisited." *Contemporary Security Policy* 22 (2): 96–120.

——. 2008. "The Israel Defense Forces in the Second Lebanon War: Why the Poor Performance?" *Journal of Strategic Studies* 31 (1): 3–40.

——. 2011. "The Rise and Fall of Israeli Operational Art: 1948–2008." In *The Evolution of Operational Art: From Napoleon to the Present*, edited by John Andreas Olsen and Martin Creveld, 137–66. Oxford, UK: Oxford University Press.

Levi, Ram. 2011. "The Fifth Fighting Space." *Israel Defense*, December 16.

Levite, Ariel. 1989. *Offense and Defense in Israeli Military Doctrine.* Jerusalem, Israel: Westview Press for Jaffee Center for Strategic Studies.

Levran, Aharon. 1990. "Threats Facing Israel From Surface-to-Surface Missiles." *IDF Journal* (19): 37–44.

Luttwak, Edward, and Dan Horowitz. 1975. *The Israeli Army.* New York, NY: Harper & Row Publishers.

Matthews, Matt. 2006. *We Were Caught Unprepared: The 2006 Hezbollah-Israeli War.* The Long War Series Occasional Paper 26. Fort Leavenworth, KS: Combat Studies Institute Press, 1–105.

Naveh, Shimon. 1996. "The Cult of the Offensive Preemption and Future Challenges for Israeli Operational Thought." In *Between War and Peace: Dilemmas of Israeli Security*, edited by Efraim Karsh, 168–87. London, UK: Frank Cass.

——. 1997. *In Pursuit of Military Excellence: The Evolution of Operational Theory.* London, UK: Frank Cass.

Opall-Rome, Barbara. 2010. "Israel Blurs Roles, Missions in Ground War Concept." *Defense News*, October 25.

———. 2014. "Israel Claims \$3b in Cyber Exports; 2nd only to US." *Defense News*, June 20.

O'Sullivan, Arieh. 2004. "Mofaz to Receive IDF's Revised Battle Doctrine." *Jerusalem Post*, March 14.

Pedatzur, Reuven. 1991. "Updating Israel's Military Doctrine." *IDF Journal* (22): 32–35.

Peres, Shimon. 1970. *David's Sling*. London, UK: Weidenfeld & Nicolson Publishing.

Posen, Barry, and Yoav Ben-Horin. 1981. *Israel's Strategic Doctrine*. Santa Monica, CA: RAND.

Rodman, David. 2000. "The Role of Israel Air Force in the Operational Doctrine of the Israel Defense Forces: Continuity and Change." *Chronicles Online Journal*, June 29.

———. 2001. "Israel's National Security Doctrine: An Introductory Overview." *Middle East Review of International Affairs* 5 (3): 71–86.

———. 2002. "Combined Arms Warfare in the Israel Defense Forces." *Defense Studies* 2 (1): 109–26.

Rosen, Stephen Peter. 1977. "A Stable System of Mutual Nuclear Deterrence in the Arab–Israeli Conflict." *American Political Science Review* 71 (4): 1367–2383.

Rothenberg, Gunther Erich. 1979. *The Anatomy of the Israeli Army: The Israel Defence Force, 1948–78*. New York, NY: Hippocrene Books.

Sanders, Ralph. 2002. "UAVs: An Israeli Military Innovation." *Joint Forces Quarterly* (33): 114–8.

Sanger, David. 2013. *Confront and Conceal: Obama's Secret Wars and Surprising Use of American Power*. New York, NY: Broadway Books.

Schiff, Ze'ev, and Ehud Ya'ari. 1984. *Israel's Lebanon War*. New York, NY: Simon & Schuster Publishers.

Steinberg, Gerald. 1990. "The Middle East in the Missile Age." *IDF Journal* (19): 30–36.

Tabansky, Lior. 2013. "Cyberdefense Policy of Israel: Evolving Threats and Responses." *Chaire de Cyber-Défense et Cyber-Sécurité* 3 (12): 1–7.

Tal, Israel. 1977. "Israel's Doctrine of National Security: Background and Dynamics." *Jerusalem Quarterly* 4 (44): 44–57.

———. 1989. "The Offensive and Defensive in Israel's Campaigns." *Jerusalem Quarterly* (51): 41–47.

———. 2000. *National Security: The Israeli Experience*. Westport, CT: Praeger Publishing.

Tamari, Dov. 2001. "Military Operations in Urban Environments: The Case of Lebanon 1982." In *Soldiers in Cities: Military Operations on Urban Terrain*, edited by Michael Desch, 29–57. Carlisle, PA: Strategic Studies Institute.

Tomes, Robert. 2004. *Military Innovation and the Origins of the American Revolution in Military Affairs*. PhD Dissertation. Baltimore, MD: University of Maryland.

Tzalel, Moshe. 2000. *From Ice-Breaker to Missile Boat: The Evolution of Israel's Naval Strategy*. Westport, CT: Greenwood Press.

Vego, Milan. 2009. "A Case Against Systemic Operational Design." *Joint Forces Quarterly* (53): 69–75.

Wald, Emanuel. 1992. *The Wald Report: The Decline of Israeli National Security Since 1967*. Boulder, CO: Westview Press.

Weinraub, Yehuda. 1988a. "Interview With Maj. Gen. Amir Drori." *IDF Journal* 3 (3): 16–20.

———. 1988b. "The Israel Air Force and the AirLand Battle." *IDF Journal* 3 (3): 22–30.

Weizman, Eyal. 2006a. "Israeli Military Using Post-Structuralism as Operational Theory." Paper presented at Conference "Beyond BioPolitics" at City University of New York, New York, NY, March 17.

——. 2006b. "The Art of War: Deleuze, Guattari, Debord and the Israeli Defense Force." *Frieze Magazine* August 3.

Yaari, Yedidia, and Haim Assa. 2005. "Dynamic Molecules: The Theory of Diffused Warfare." *Pointer – Journal of Singapore Armed Forces* 31 (3): 5–16.

Zisser, Eyal. 2004. "The 1982 'Peace for Galilee' War: Looking Back in Anger – Between an Option of a War and a War of No Option." In *Never-Ending Conflict: Israeli Military History*, edited by Mordechai Bar-On, 193–211. Westport, CT: Praeger.

4 Searching for security, autonomy, and independence[1]

The challenge of military innovation in the ROK armed forces

Since the end of the Korean War, which ended in an armistice agreement signed on July 27, 1953, South Korea/ROK's security has been in unnatural conditions with the continuous presence of preparation for war or the serious expectation of war – a condition referred to as unstable peace (Boulding 1978, 13). Indeed, for more than six decades, the country has faced an array of persistent security dilemmas brought out by the confluence of its geostrategic location, historical legacies, powerful national sentiments, alliance politics, and superpower rivalries. The divided Korean Peninsula has been at the center of complex power constellations in Northeast Asia that included historical fault lines of prolonged hostilities, mutual suspicions, territorial disputes, and geostrategic competition between the United States, Soviet Union, China, and Japan (Godement 2005; Scalapino 2010). Its strategic importance has been tied to the regional balance of power politics and amplified by the presence of four of the world's six largest militaries and three recognized nuclear powers (China, Russia, and the United States), not including the North Korea. Notwithstanding its economic transformation on the global stage over the last three decades, one could argue that in relation to its respective neighbors – by the degree to which the strength at its disposal matched its national goals and ambitions – South Korea resembled a small state in the relational context, defined by its historical and geostrategic predicaments, patterns of behavior in the international system, and conditions of relative strategic inferiority.

The prevailing sense of insecurity emanating from the complex matrix of larger systemic factors – superpower involvement with contending national interests – has shaped South Korea's threat perceptions, foreign policy options and choices, as well as strategic culture (Cho 1989; Bae and Moon 2005). Yet, by far, the defining security predicament for South Korea has been embedded in the nearly zero-sum conflict and profound rivalry with North Korea/DPRK over political legitimacy, territorial sovereignty, and strategic orientation. In particular, South Korea's security paradigm has for nearly 70 years focused on sustaining the status quo – maintaining deterrence and robust defense posture in order to prevent another major outbreak of war on the Korean Peninsula (Lee 2005a, 13). Its three mutually reinforcing strategic pillars – (1) defensive deterrence, (2) U.S.–ROK alliance, and (3) forward active defense – have defined the baseline of South Korea's national security conceptions as well as corresponding force structure and operational conduct (Moon 2000, 96–99). Regardless of the systemic changes in the global and regional security environment, this fundamental strategic raison d'être and principal objective has remained relatively constant. From a South Korean perspective, as long as North Korea retains an independent capability for conducting an all-out conventional offensive into the ROK (by design, accident, or default) aiming to either inflict a substantial damage or reunify the Korean Peninsula on its terms, South Korean defense planners cannot fully discard its validity (ROK Ministry of Defense 1998). Moreover, with the absence of a permanent peace treaty, robust military deployments on both sides of the Demilitarized Zone (DMZ) that separates both North and South Korea, and continued regional and superpower involvement, security on the Korean Peninsula remains relatively an elusive concept bound by a number of uncertainties and risks – a deeply embedded predicament that continues nearly three decades after the end of the Cold War.

However, since the early 1990s, South Korea's security dilemmas have become more "fluid" and multifaceted with broad implications in both strategic and operational domains. There are at least five key factors that have redefined South Korea's security equation over the last decade: (1) the expansion of the North Korean threat spectrum, (2) shifts in the U.S. global and regional military strategies, (3) the rise of China, (4) South Korea's aim for greater autonomy resulting from its improved economic and military capabilities, and (5) subsequent changes in the U.S.–ROK alliance (Klinger 2008). To begin with, notwithstanding the prevailing conventional threats (i.e., scenarios and contingencies linked to low- to high-intensity conventional wars), South Korea's conflict spectrum over the past decade has increasingly faced threat assessments that include a range of asymmetric and non-linear security challenges. On the one end of the threat spectrum is North Korea's continuously advancing ballistic missile program coupled with its WMD (nuclear, chemical, and biological) development. North Korea has recalibrated its strategies toward WMD development and testing in order to offset the U.S.–ROK conventional military superiority, attain greater economic and political leverage, and, perhaps most importantly, guarantee its regime's survival. According to the Korea Institute for Defense Analyses, North Korea currently possesses the world's third largest arsenal of chemical and biological weapons and the sixth largest arsenal of missiles (Kim 2009). Its nuclear weapons capability, demonstrated by three underground nuclear weapons tests in 2006, 2009, and 2013 coupled with diverse inventory of ballistic missiles, has served as a major asymmetric force multiplier to North Korea's internal economic, technological, and military deficiencies (Lee 2001).

On the other end of the threat spectrum, however, is North Korea's specter of a failed state – its gradual socioeconomic decay accompanied by internal structural erosion and prolonged international diplomatic isolation, which have broadened the risks of potential instability and volatility – that is, scenarios ranging from North Korean implosion, collapse, and attendant external involvement in Korean unification modalities (Lee and Pollack 1999). Although the country has endured its abysmal socioeconomic decline for more than a decade, largely by exploiting external aid, its "military-first policy" at the expense of economic reforms has further deepened the country's pervasive economic crisis. Amid these "systemic" transformations in the source, nature, and character of North Korean security challenges, South Korean defense planners have been constrained by the risks and costs of potential confrontations, spillovers, or crises (Scobell and Sanford 2007). In other words, one could argue that "[while] the 'vertical' threat spectrum has decreased significantly since the early 1990s, the 'horizontal' threat spectrum – the operational convergence of asymmetrical capabilities on the part of the North coupled with growing possibility of non-linear scenarios has actually increased" (Lee 2005b, 254). Notwithstanding the progressive complexities in North Korea's "hybrid" conflict dynamics – an amalgam of diverse security challenges and threats – South Korea's defense planning has been also increasingly taking into account regional strategic drivers, military modernization drives, and sharper power projection capabilities of its neighbors, particularly China. The economic, political, and military rise of China, embedded in three decades of relentless Chinese economic growth, has propelled progressive modernization of the Chinese military with major improvements in virtually every capability domain: land, air, naval, missile, space, cyber, and electronic warfare (Tellis 2012). The cumulative effects of these developments have been substantial as China's catalogues of air, land, and naval platforms have been gradually catching up in terms of both qualitative sophistication and operational effectiveness (Erickson 2012). In March 2014, China announced $131 billion defense budget, up 12.2 percent from the 2013 budget of $119 billion – marking 17 straight years of double-digit increases in defense spending (Minnick 2014). With China expanding its national interests in the broader context of "new historic missions," it seeks to regain a great power status and reassert its geopolitical role in the region. The PLA has been pursuing capabilities and corresponding infrastructure for force projection into China's "near seas" or an area defined by the "first island chain" consisting of the Kuril Islands, Japan, Taiwan, and the South China Sea. China calls its comprehensive strategy a "counter-intervention," which is interpreted in the Western strategic thought as denying U.S. forces the freedom of action in China's "near seas" by restricting deployments of U.S. forces into theater (anti-access) and denying the freedom of movement of U.S. forces already there (area denial). In the long term, China envisions capabilities that would extend its strategic reach into the "second island chain," which includes U.S. bases on Guam (Schreer 2013, 8).

China's growing strategic imprint – whether direct or indirect – has been also increasingly tied to the security and stability of the Korean Peninsula, providing both opportunities and new challenges for the U.S.–ROK alliance. On one hand, China's geopolitical and economic rise coupled with its integration in the global community has given its diplomacy more leverage in managing tensions on the Korean Peninsula (Kim 2006). Since 2003, Beijing has been more proactive in mitigating crises by providing critical economic lifeline – energy and food aid – to North Korea while attempting to contain Pyongyang's nuclear ambitions. In this context, China's three strategic objectives toward North Korea have been traditionally interpreted in terms of achieving (1) stability (no implosion, no regime change, and no war), (2) peace (diplomatic normalization between the

United States and North Korea), and (3) denuclearization/non-proliferation of WMD on the Korean Peninsula (Snyder and See-Won 2013). Notwithstanding signs of increasing strains in the Sino-North Korean relations, following North Korea's third nuclear test on February 12, 2013, China's key strategic priority in preventing a major war on the Korean Peninsula has aimed at preventing a North Korean implosion. Such collapse would undermine China's geostrategic interests by removing traditional strategic buffer vis-à-vis the United States provided by North Korea and significantly increase PLA's military deployment requirements in northeast China. In this view, China cannot afford to cease its support, trade, and aid to North Korea. Beijing has therefore prevented North Korea's socioeconomic implosion, while simultaneously exerting pressure on Pyongyang to return to the stalled Six-Party Talks and resolving North Korea's nuclear issue through multilateral diplomatic channels.

On the other hand, China's policy toward the Korean Peninsula has been also increasingly conditioned through its perceptions and responses to the U.S. strategic rebalancing to the Asia Pacific. In particular, China has viewed U.S. rebalancing efforts increasingly contentious and in "zero-sum" terms – as a comprehensive strategy by Washington to curtail China's rise and influence in the region (Yun 2013). Beijing has periodically criticized U.S. military deployments in East Asia, along with increasing military cooperation with Japan and South Korea (Stratfor 2010). Accordingly, China has been preparing for select contingencies in North Korea, including the possibility of its collapse. China's PLA has been training to conduct three types of missions in North Korea: (1) "humanitarian missions" – assisting refugees by providing emergency food, medical, and energy supplies; (2) peacekeeping or "order-keeping missions" – internal control missions such as serving as civil police to strengthen controls in and around vital border crossings with North Korea; and (3) "environmental control missions" to minimize nuclear contamination from a potential strike on North Korean nuclear facilities, or securing nuclear weapons and fissile materials (Glaser, Snyder, and Park 2008). Consequently, China could mitigate the freedom of action of U.S.–ROK forces by disrupting, limiting, preventing, and intervening in the U.S.–ROK military responses in and around the Korean Peninsula.

East Asia's changing strategic realities over the past two decades have had significant ramifications on South Korea's security conceptions and force modernization trajectory. In particular, South Korean defense planners realized the need to maintain its traditional collective defense mechanism and robust force posture vis-a-vis North Korea but at the same time also addressing security issues that have a strong Chinese imprint. Simultaneously, the changing security dynamics on the Korean Peninsula has accelerated changes in the U.S.–ROK alliance as well as shaped the path and patterns of South Korea's defense reforms. Since the late 1990s, when the ROK Army (ROKA), Navy, and Air Force first published their conceptual visions of future warfare, South Korea has been rethinking its defense requirements while searching for relevant strategic, organizational, and operational concepts that would address existing as well as future-oriented defense requirements. In other words, South Korea's defense planners have been searching for a new strategic paradigm that would allow greater flexibility, adaptability, and autonomy under conditions of strategic uncertainty. In doing so, however, South Korea has been caught between political, historical strategic legacies, emerging complex threats, interoperability requirements and linkages with the U.S.–ROK alliance that have precluded significant defense transformation. Indeed, the progress in translating South Korea's RMA strategic visions and operational concepts into practice has proved challenging given a range of persisting budgetary constraints, institutional barriers to change, and the prevalence of traditional strategic culture. Accordingly, South Korea's paths and patterns of military innovation must be therefore linked to the broader context of its national security debate on the future direction and character of its strategic posture, national defense strategies, and the continuity and change in the U.S.–ROK alliance.

Based on the above argument, this chapter provides an overview of South Korea's search for military innovation. To do so, the chapter first outlines the baseline assumptions underlying South Korea's traditional security paradigm, its conditioning factors, and three main elements – (1) defensive deterrence, (2) U.S.–ROK alliance, and (3) forward active defense. It also outlines the basic patterns in the U.S.–ROK combined operational planning, training, and conduct. The second part of the chapter then contextualizes the "Korean" RMA conceptual adaptation and debate based on the various internal and external perspectives in two major phases of the 1990s and 2000s. The chapter combines existing open-access literature pertaining to the key issues and factors shaping South Korea's strategic dynamics supported by interviews with select Korean military, government officials, and leading security experts.

Revisiting South Korea's traditional security conceptions

Since October 1, 1953, South Korea has defined its security through a bilateral collective defense framework of the U.S.–ROK alliance, codified in the 1953 Mutual Defense Treaty, which provides assurances of defense and deterrence vis-à-vis external military threats. In this context, South Korea's traditional security paradigm centered on defending the ROK from any potential North Korean attack or provocation and has been inherently tied to U.S. threat perceptions, national security interests, doctrine, command structure, and force deployment in East Asia (Lee 2005a). The U.S. Army personnel provided strategic guidelines and essentially shaped the joint command and planning process, doctrine, as well as the training of South Korean forces. Until the early 1980s, the ROKA has been organized and trained based on the U.S. doctrinal lines of the early 1950s, which emphasized "attrition" warfare and firepower. From the mid-1980s, the ROK forces adopted select concepts of the U.S. ALB doctrine. Nearly all equipment of the ROK forces was of U.S. origin or design, and its tactical concepts and force deployment directly patterned the U.S. Army, Navy, and Air Force (Luttwak and Canby 1983). Many South Korean officers and senior staff enlisted receive training at U.S. Army schools and institutions, and U.S. officers teach at ROK military academies. Accordingly, one could argue that there were no distinct "Korean" strategic or operational concepts per se – South Korean defense plans depended almost entirely upon the United States, which provided security assistance and assurances through its foreign military sales, intelligence cooperation, combined training, and, most importantly, through its extended nuclear deterrence and direct military presence in South Korea (McLaurin and Moon 1989). That said, however, the bulk of the defense strength in terms of manpower has been arguably provided by the South Korean forces – primarily the ROKA supported by independent air force and a navy (complete with ROK Marine Corps).

The U.S.–ROK defense planning has traditionally recognized the importance of three key constraints in defending South Korea: (1) *geographic*, (2) *asymmetric*, and (3) *political* (Luttwak and Canby 1983). First, South Korea lacks strategic depth, which essentially precludes any type of elastic defense (i.e., defense in depth that trades space for time) and limits early-warning options. In geographical terms, the distance between the DMZ and Seoul – the political, business, and cultural center of South Korea – is only approximately 40 km, making the densely populated capital city with over 11 million inhabitants highly vulnerable to a North Korean ground or artillery attack (U.S. Department of Defense 2000). Moreover, the highly mountainous terrain and complex topography around Seoul poses constraints to maneuverability, targeting, communications, logistics, and fires (Hollis 1993). Cross-country movement by wheeled and tracked vehicles is difficult in the narrow mountain areas, which constitute nearly 70 percent of the Korean Peninsula. In this setting, the extremely small but highly populated combat radius around Seoul amplifies the risks of high collateral damage and major socioeconomic disruptions in any type of crises or conflict scenarios – ranging from low-intensity contingencies that include long-range artillery fire, limited missile strikes, and special forces incursions, to full-scale high-intensity assaults (Bennet 1993). Accordingly, any limited operations by either side could effectively trigger uncontrollable or unintended escalation (Collins 1994, 21). These geographic predicaments have made a forward defense of Seoul as a military and political imperative – essential to the survival of South Korea.

The second key conditioning factor imposing fundamental constraints in the U.S.–ROK defense planning has been the quantitative asymmetry, disposition, doctrinal orientation, and deployment of North and South Koreas' armed forces. In particular, North Korea's conventional forces have enjoyed relatively significant numerical superiority over South Korea in terms of manpower, armor, and artillery equipment. Notwithstanding its prolonged economic hardships, supply shortages, lack of new equipment, and declining readiness amid deteriorating internal conditions over the past two decades, North Korea has been able to sustain and even expand its conventional forces to the fourth largest in the world – with over 1.1 million Korean People's Army (KPA) personnel in active service (compared to South Korea's 687,000) and a reserve force totaling nearly 5 million (IISS 2010, 394). North Korea's military grew from 420,000 troops in 1972 to 750,000 in the 1980s, 1 million in the 1990s, to 1.1 million in 2010 (Kim 2010). The KPA has also maintained the largest special operations force in the world totaling about 100,000 highly trained "commando" troops whose missions include reconnaissance, attack or seizure of critical positions, and interference of South Korea's defenses (Hodge 2003, 73). While North Korea's quantitative advantages have been gradually offset by the U.S.–ROK conventional superiority including intelligence, air power, command and control systems, education, and training, the sheer quantity of North Korean infantry divisions, supported by diversified array of armor, heavy artillery, and special forces concentrated in forward lines near the DMZ, has inherently created "a quality on its own" (Kim 2007). Accordingly, while the age and obsolescence of many North Korean combat systems coupled with the lower training hours of their crews cannot match U.S.–ROK capabilities, their potency and risks to inflict significant damage or launch selective or massive conventional attacks against South Korea should not be underestimated.

The quantitative asymmetry between North and South Korea's forces has been also amplified by the differences in their doctrinal and political orientation. Taking into account its periodic doctrinal adjustments that emulated select elements of the Soviet and Chinese doctrines, North Korea's military strategy and operational doctrine has traditionally emphasized the primacy of the offence that would provide a military option for reunification of the Korean Peninsula by force (Hodge 2003, 69). Indeed, North Korea's military doctrine has evolved through at least four stages since the founding of the KPA in 1948. These include: (1) Doctrine of Regular Warfare (1951–1962) based on Soviet military doctrine and operation art modified on the basis of the Korean War experience; (2) Doctrine of People's War (1962–1976) based on Maoist protracted war of attrition, mobilization of the entire population, and political and ideological dimensions of warfare; (3) Doctrine of Modern War (1976–1990) based on the quality of arms and level of military technologies with emphasis on firepower; (4) Modern Combined Arms Doctrine (1990s onward) based on firepower and mobility as a solution (Minnich 2005, 73–78). North Korea's "Five-to-Seven" strategic plans – dating back to the 1980s – envisioned reaching Seoul in five days and the rest of South Korea in seven days (Kim 2010). To do so, the KPA's traditional operational concepts have been based on fighting a *quick decisive war* with a mix of regular combat forces and special operation forces that would conduct a series of *surprise attacks* along the major invasion corridors leading to Seoul (i.e., Munsan, Chorwan, and Tongduchon approaches) as well as in the rear areas of the ROK. The combined employment of KPA forces, called *mixed tactics*, would be aimed at achieving three objectives: (1) breaching the defenses along the DMZ, confusing the enemy's command, disrupting and destroying the first echelons of the U.S.–ROK forces; (2) isolating Seoul, securing terrain, and consolidating gains; and (3) pursuing and destroying the remaining forces throughout the depths of the battlefield, occupying the remainder of the peninsula, and defending North Korea – ideally prior to the ROK mobilization and the arrival of U.S. reinforcements (Minnich 2005, 67). In this context, the KPA's surprise attack strategy or "blitzkrieg version" relies on mutually reinforcing elements of *mass and dispersion* (concentrating combat power at the decisive time and place), *surprise* (i.e., weather, timing, covert infiltrations), *maneuverability* (operational and tactical mobility by employing both ground and air transportation assets to quickly maneuver infantry, artillery, and armor on the battlefield), overwhelming *firepower*, and *deception strategies* (Minnich 2005, 75–78). The offensive character of North Korea's military strategy can be seen in its force deployment – with nearly two-thirds of the KPA's ground forces and their logistical support (including underground tunnels) concentrated in the forward areas between Pyongyang and the DMZ. This disposition inherently limits South Korea's early warning time and provides the initial advantage to North Korea with its ability to mass forces vertically in multiple corridors against South Korea (Lee 1988).

The combined effect of both asymmetric and geographic constraints has essentially shaped the third and perhaps most important category of conditioning factors that defined South Korea's geostrategic inferiority, which are *political* (Luttwak and Canby 1983). Namely, five key political constraints shaped the traditional South Korean security paradigm: (1) the difficulties in ascertaining North Korea's intentions, threat estimates, and politico-military strategies, particularly amid its gradually deteriorating socioeconomic conditions, which has amplified security uncertainties and risks of potential miscalculation, and simultaneously mitigated the reliability of South Korea's deterrence; (2) the excessive costs, sometimes referred to as "strategic premium" of any preemptive actions or military initiative by South Korea or the United States in striking at North Korean force concentrations before they could counterattack; (3) the prohibitively high socioeconomic costs of building up a preclusive type of defense or a force strong enough to defeat North Korea's invasion at the onset, before serious damage could be inflicted on the Seoul area; (4) the continuous superpower involvement on the Korean Peninsula, particularly during the Cold War, dictated by the logic of bipolarity and extended deterrence that either enabled or constrained both North and South Korea's foreign policies and defense strategies; and, in this regard, (5) the ambivalence and political ramifications of de facto security dependence on the United States (at least until late 1970s), through which South Korea was inherently subjected to various economic pressures and political involvement by the United States in exchange for U.S. security assistance and bilateral security cooperation (McLaurin and Moon 1989).

Three pillars of South Korea's defense strategy

The above conditioning factors have subsequently defined South Korea's defense strategy and its three key pillars: (1) defensive deterrence, (2) U.S.–ROK alliance, and (3) forward active defense. While their core characteristics have not been codified or explicitly presented in any official "Korean" strategic doctrines, its contours can be extrapolated from analyses of ROK defense white papers and a range of studies concomitant with South Korea's national security issues and defense planning dynamics during and post-Cold War (Clough 1976; Johnson and Yager 1979; Curtis and Han 1983; Lee, McLaurin, and Moon 1988; Olsen 1988; Simon 1993; Blackwill and Dibb 2000). The first pillar underscoring South Korea's traditional defense strategy – defensive deterrence – emphasizes two objectives: deterring military threats in peacetime and defending South Korea in wartime (Han 1999). In other words, the strategy implies preventing a major war by effectively deterring North Korea-initiated military provocations and, if deterrence fails, winning the war. According to the 1994 ROK defense white paper, "the enemy's will to provoke war can be undermined only when our own stance shows clearly that we are capable of and will employ an effective deterrence in times of contingency" (ROK Ministry of National Defense [MND] 1995, 84). Defensive deterrence thus stipulates the need for a "visible combat capability that would deter any offensive military moves by North Korea" (Moon 2000, 96). In the context of South Korea's traditional security paradigm, visible combat capability essentially means the primacy of conventional ground forces – the ROKA, which has accounted for the majority of South Korea's total military manpower and dominated strategic and operational planning, resource allocation, and force structure. The ROKA has trained to counter a potential North Korean attack immediately by blocking KPA's first echelon forces, defending the capital and rear areas, and executing coordinated air-ground attacks from the initial stages of war – prior to the arrival of U.S. reinforcements (Han 1999).

In order to amplify a credible defensive deterrent posture, the second element in South Korea's traditional military strategy has focused on effective external management, coalition-style warfare, and bilateral alliance with the United States. According to Chung-in Moon, "if South Korea was to survive in the tough security environment, effective external management was as critical as military self-help because of the country's inherent military weakness ... military deterrence through the alliance with the U.S., therefore, was the backbone of South Korea's external security management during the Cold War, limiting the utility of other alternatives" (1998, 270). Codified in the framework of the 1953 U.S.–ROK Mutual Defense Treaty, operationalized through the U.S.–ROK Combined Forces Command (CFC), created in 1978, and, perhaps more importantly, solidified through the tangible presence of U.S. forces in South Korea (USFK – 2nd Infantry Division, 8th Army, 7th Air Force, and an array of C3I networks), the U.S.–ROK security cooperation has served as "the fundamental axis for security on the Korean Peninsula" (ROK MND 1996, 108). Any major scenario of North Korea's military aggression would automatically activate U.S. involvement ("trip wire" analogy) in three ways: (1) by assuming wartime operational command and control (OPCON) of all South Korean forces; (2) by engaging its forward-deployed combat forces and their capabilities in immediate defense of South Korea; and (3) by providing massive reinforcements – primarily command and control, strategic surveillance, and naval and air defense assets – and initiating air and sea lifts from the U.S. military bases in Japan as well as the continental United States (Moon 2000, 97). In this context, the U.S.–ROK alliance has shaped nearly all aspects of South Korean strategic and military doctrine, defense planning and management, training and exercises. The principal objective has been to continuously enhance the interoperability, coordination, and compatibility of U.S.–ROK weapons platforms, C4ISR systems, combined training, and organizational structures. Overall, for nearly six decades the alliance has inherently enabled the United States to maintain "a critical strategic footprint on the Asian continent throughout the Cold

War, while at the same time South Korea was able to fully exploit the benefits arising from the U.S. security umbrella" (Lee 2005a, 16).

The third pillar of South Korea's traditional defense strategy is the operational concept of "forward-active defense" (FAD). Until the early 1970s, U.S.–ROK defense planners worked on phased withdrawal strategies as a response to a potential North Korean attack. This strategy assumed that in case of a full North Korean invasion, envisioned as a sustained, all-out offensive similar to the 1950 invasion, South Korea would absorb the momentum of a North Korean invasion, trading space (Seoul) for time, regroup, and counterattack in superior strength with large-scale reinforcements from the continental United States. This strategy was similar to the U.S.-style attrition defense and assumed that North Korea would have to concentrate its forces in specific areas in order to attack South Korea's defense lines, which would increase their vulnerability to massed artillery fire and air attacks. Under this logic, the invading KPA echelons would exhaust their supply lines, mainly due to continuous allied air interdiction along the primary invasion corridors, while the U.S.–ROK forces would fully mobilize in the rear areas (Lee 1988). However, from the mid-1970s, when North Korea embarked on a comprehensive military build-up, it became increasingly clear that its frontal assault strategies and operational concepts shifted toward fighting a "blitzkrieg"-type warfare (KPA's "Doctrine of Modern War") – brief, rapid, high-intensity maneuver operations conducted by both infantry and special operations forces, aiming to capture Seoul in a matter of days, while launching diversionary attacks on ROK's strategic assets in the rear areas. The close-in nature of fighting around Seoul and the limited early warning would likely offset the pace and effectiveness of the phased withdrawal and increase the risks of U.S. reinforcements not being able to arrive in South Korea in time (Pitt 1987). Moreover, under phased withdrawal, there would be concerns about the possible collapse of morale of the U.S.–ROK forces as well as public confidence (Luttwak and Canby 1983). As John Collins noted, "forced evacuation of that immense city [Seoul] would cause vast human misery. A protracted war up and down the peninsula in a worst case scenario would destroy the infrastructure that South Korea has so painstakingly constructer over the last four decades, shatter its economy, and fundamentally disrupt every facet of national life" (1994, 16). For these reasons, U.S.–ROK defense planners revised the original plans in the mid-1970s and shifted their focus on sustaining defense position on the forward line north of the Han River dividing Seoul (Lee 1988). Dubbed the "Hollingsworth Line" after Gen. James F. Hollingsworth, Commander of the USFK and CFC (1973–1976), it embodied three defensive belts, so-called *Forward Edge of the Battle Areas* (FEBA – Alpha, Bravo, Charlie), that would thwart the advancing KPA echelons (mainly along the Kaesong–Munsan invasion corridor) from overtaking Seoul in the early stages of hostilities. The early versions of the plan developed by the CFC, OPLAN 5027–74, envisioned eight of the 19 ROK infantry divisions covering the first belt (FEBA Alpha) and utilizing most of allied combat power (i.e., artillery, tanks, infantry, and defensive fortifications) situated in the Military Control Zone that runs five miles south of the DMZ. The rest of the forces would be deployed along the second and third belts

(FEBA Bravo and Charlie) about 50 miles from the DMZ and beyond the range of the bulk of North Korea's artillery (Suh 1999). From this position, the allied forces would prepare for U.S. reinforcements and subsequently conduct retaliatory strikes (Roehrig 2006). In order to reinforce the credibility and reliability of the U.S.–ROK deterrent posture, Hollingsworth then proposed to transfer the war to North Korea's territory, seize the North Korean staging city of Kaesong, conduct intense air strikes, and, ultimately, capture the capital of Pyongyang within nine days. The U.S.–ROK forces would create a reverse asymmetry by maximizing the cumulative attrition of the enemy forces, occupy the north, seize its leadership, and, ultimately, win the war on South Korea's terms. This concept became known as "forward-active defense" and subsequently shaped the direction of subsequent U.S.–ROK operational war plans. According to Moon, underscoring the FAD is the assumption that "any failure of deterrence will be linked automatically to winning the war and occupying the North, which [would end North Korea's regime] and expedite the process of unification" (2000, 97).

With the introduction of the U.S. ALB doctrine in 1982, the operational concept behind FAD was revised further – particularly in the context of understanding the battlefield with regard to space, time, and depth (Pullan 2005). The idea was to change the traditional linear understanding of battle – by extending the battlefield space beyond the immediate forward lines and focusing on the entire depth in order to delay, disrupt, and destroy North Korea's uncommitted forces and second and third echelons. In this context, the ALB in Korea would be conceptualized as an integrated battle of forward defense and deep attack – in conjunction with the protection of the rear areas (Lee 1988). This approach emphasized attacks in depth, operational level thinking, rapid indirect approach, synchronization, and the use of advanced weapons systems as force multipliers. Specifically, by blurring the front and rear areas and unifying air and ground operations, U.S.–ROK forces would seize and hold the initiative across the battlefield (rather than just front lines) and essentially "throw the enemy off balance" by striking the KPA from unexpected directions or dimensions. This would prevent the KPA's follow-on echelons to exploit advantages of the initial assault. The rationale was to avoid costly sequential battles of attrition and offset the growing quantitative asymmetries in force ratios that favored North Korea (Lee 1988).

Throughout the 1990s and 2000s, the U.S.–ROK forces periodically revised their operational concepts, based on changes in North Korea's force structure and deployment, as well as doctrinal shifts and lessons learned in the U.S. military – that is, the introduction of the U.S. Army's *Field Manual 100-5* in 1993. Internal CFC publications such as *the Deep Operations Primer-Korea*; *Air Ground Operations-Korea*; *Joint/Combined Fires-Korea* have been essentially based on select U.S. operational concepts (i.e., "Deep Battle," "Joint Fires," "Network-Centric Warfare") and adjusted to the Korean Peninsula (Combined Forces Command 1995, 2002). For example, the USFK developed a concept called "Deep Battle Synchronization Line" to synchronize and coordinate deep operations with joint fires, which has been very similar to the coordination requirements and processes for employing fires outlined in the U.S. Joint Operations doctrine 3.0 (D'Amico

1999, 72). In order to ensure doctrinal compatibility, operational interoperability, and combined defense capabilities with U.S. forces, South Korea's military doctrine inherently emulated U.S. doctrinal templates and concepts. In 1993, for example, Commander-in-Chief of the CFC, Gen. (ret.) Robert RisCassi, noted, "there are no U.S. or ROK sectors – only a combined battlefield ... Because of the nature of the North Korean threat and the terrain upon which a war would be fought, South Korea's combined defense must be seamless. Korean and American units must rely on each other for too many battlefield functions to allow national divisions to artificially separate one from another ... Unless we talk the same technical language, there will be a great deal of unwelcome friction" (Hollis 1993, 7–8). In this context, the U.S.–ROK forces have trained combined air, naval, and ground operations, strengthening capabilities that would achieve prescribed political and military objectives in diverse North Korean conflict scenarios.

Aiming for "self-reliant defense"

For more than five decades, the above strategic conceptions of conventional deterrence and defense have ensured at least partial strategic stability and status quo on the Korean Peninsula. Notwithstanding the diverging types of military provocations, crises, and intrusion by North Korea throughout the Cold War, U.S.–ROK political and strategic considerations have largely refrained from more active and direct responses in order to prevent a major escalation, superpower intervention, and another fratricidal Korean conflict. As Gen. (ret.) Jae-Chang Kim of ROKA noted, "During that period, patience was a virtue for the [U.S.–ROK] alliance" (2003, 59). With the end of the Cold War, however, the security dynamics on the Korean Peninsula have experienced dramatic shifts resulting from a confluence of external and internal drivers at the strategic, operational, and tactical levels that have altered both North and South Koreas' security equation. To begin with, the collapse of the socialist Soviet bloc has brought comprehensive political and economic isolation for North Korea's regime, which became paralyzed in its inability and unwillingness to adopt meaningful politico-economic and societal reforms without endangering the prospects of its survival and raison d'être. North Korea's economy has essentially faltered in the 1970s, declined in the 1980s, and collapsed in the 1990s (Oh and Hassig 1999). In the absence of viable alternative political options and unable to pursue broad socioeconomic reforms, North Korea under Kim Jong-il resorted to strengthening its "military-first" policy orientation. This strategy emphasized channeling North Korea's limited resources into the development and production of asymmetric forms of warfare such as long-range ballistic missiles, WMDs, special forces, and other asymmetrical force upgrades to strengthen the internal legitimacy and survival of the regime, while providing external leverage in extracting political concessions and foreign aid from the international community. Consequently, the accelerated acquisition and testing of asymmetrical systems by North Korea under Kim Jong-un has undermined U.S.–ROK conventional deterrence. In the increasingly "hybrid conflict" spectrum, U.S.–ROK forces would have to prepare not only for conventional all-out war scenarios (i.e., responding to North Korea's surprise attack as quickly as possible; defending the capital and rear areas; and defending territorial sea and air simultaneously) but also provide effective defense against possible long-range artillery, ballistic missile attacks, rear-area infiltrations, and other forms of low-intensity and asymmetric forms of warfare. The widening operational requirements over the last two decades have thus propelled the rationale in South Korean defense establishment to pursue RMA-oriented comprehensive force modernization of ROK forces in areas of air and naval power, strategic and tactical surveillance, early warning, C4I, battle management, target acquisition, stand-off precision strike, and network-centric warfare.

In this context, the diffusion of RMA-oriented concepts and technologies in South Korea's military modernization has progressed on two parallel trajectories: (1) U.S.–ROK combined forces dimension: shaped by the changes in the character of North Korean threats as well as shifts in U.S defense strategies and operational conduct; and (2) Korean/internal RMA debate in the context of its strategic assessments, defense reforms, and operational requirements for "self-reliant national defense" – aimed at attaining more independent and technically advanced operational capabilities. Inherently, one could argue that the former has essentially shaped the latter by providing a baseline for defense planning, interoperability requirements, and training of the ROK military. In other words, select U.S. RMA-oriented defense transformation concepts – that is, information superiority, precision strike, battlefield situational awareness, network-centric warfare – have gradually permeated into the U.S.–ROK combined training and operations and subsequently provided guidelines for South Korea's own military modernization concepts, including its doctrinal adjustments, organizational reforms, and defense-technology development, acquisition, and procurement. As Chang-Hee Nam noted, "since the ROK military has long wished to enhance its war fighting effectiveness by employing the NCW and EBOs concepts, the USFK's transformation propelled the ROK to proactively embrace reform" (2007, 172). Indeed, for more than a decade, the ROK military has attempted to conceptualize and synchronize its own distinct "Korean RMA" by emulating and adapting select U.S. defense transformation concepts. This parallel and evolving "Korean RMA conceptual trajectory" can be seen in South Korea's efforts to enhance its independent strategic planning, interoperability with the U.S. forces, and, more importantly, its broader aim to achieve "self-reliant" military capabilities.

Acknowledging that South Korea's security has been overly dependent on the U.S.–ROK alliance, the country's political and military elite have consistently voiced the need for a comprehensive ROK military modernization that would amplify South Korea's deterrence and defense capabilities vis-à-vis North Korea as well as for long-term unspecified threats. Yet, the compelling and relatively ambitious character of Korean RMA visions of the past two decades have been in sharp contrast to the prevailing structural, political, and political realities that have sustained traditional institutional resistance to change. Notwithstanding concerted efforts to create smaller but smarter "advanced elite force," Korean RMA drive has not fully eradicated the power of the "old" paradigm, which has been sustained by the confluence of strategic legacies, economic and budgetary constraints, and political conservatism that has inherently precluded a greater flexibility and adaptability in pursuing viable defense reforms. However, the ROK military and its service-specific components have acknowledged the need to transform their traditional linear defense planning, force structure, strategic and operational concepts by adopting specific RMA-oriented concepts and technologies relevant to Korean strategic context. Seen from this perspective, the next section attempts to construct a historical trajectory of Korean RMA path of emulation and adaptation based on concomitant debates that defined its military modernization during the 1990s and 2000s. While by no means complete, the review sketches how the RMA diffused in South Korea, mapping its conceptual thrust in line with policy adaptation and implementation challenges. In doing so, it identifies two progressive phases: (1) ROK's future battlefield concepts in the 1990s, and (2) U.S.– ROK alliance transformation of the 2000s.

Future battlefield concepts in the 1990s

Since the early 1990s, the ROK MND, ROK Armed Forces, and a small number of Korean/U.S. defense analysts began to conceptualize long-term force modernization visions based on the evolving strategic priorities and increasing scope of defense requirements, which propelled the need for new thinking, innovation, and military reforms beyond the traditional defense paradigm. For example, in 1991, the MND published a report titled *Defense Policy Development Trends in the 1990s* followed by a 1992 document titled *Defense Policy toward the 21st Century*. In 1993, the MND published its *Mid-Term Defense Policy Report*, and in 1995, the MND completed its first version of a comprehensive defense policy review titled *Basic Defense Policy Report* (Lee 2000). These reports emphasized the need to streamline the ROK armed forces, enhance defense management, and build more technologically oriented force in order to adjust to the changes in global and regional security environment in the post-Cold War era (Cha 1994; Pollack and Cha 1994). In retrospect, these reports also reflected early attempts to redefine South Korea's strategic outlook and options based on the shifting contours of major power interactions, the rise of new security threats, Korean unification debate, and expanding U.S.–ROK alliance demands. However, their findings offered only limited policy choices and have not translated into adequate momentum that would change the entrenched institutional conservatism within the ROK armed forces (dominated by the ROKA), as well as a range of political constraints precluding military reforms. According to Chung Min Lee (2000), "while past attempts also emphasized the need for reforms, the implementation phase was largely neglected so that once the political will for defense reforms waned; there was no movement within the MND or the armed forces to pursue much needed reforms. Another major factor was the politicization of key reform issues such as the push for an 'Integrated Force' in the early 1990s."

From the mid-1990s onward, however, the U.S. military accelerated its IT-RMA drive, revising its doctrines, operational concepts, training, education, weapons procurement, and organizational structure in order to exploit continuing advances in information and communications technologies, surveillance and intelligence, command and control systems, and precision munitions. The resulting "RMA Technophilia" of the mid-1990s propelled the development of RMA-oriented initiatives, processes, and programs, which aimed to significantly increase U.S. military capabilities and overall combat power across a full spectrum of operations. In this context, select U.S. RMA concepts and technologies began to permeate into the Korean theater of operations and shaped the character and direction of the USFK and its training and operational conduct. Subsequently, the CFC along with the U.S. Pacific Command began to update their operational concepts and plans (i.e., OPLAN 5027-94), revising the existing forward-active defense template, while preparing for a range of alternative contingencies and scenarios in case of renewed hostilities (Hollis 1993). The widening operational spectrum included options for a naval blockade, surgical preemptive strikes against North Korea's nuclear facilities, and possible unification contingencies and scenarios concomitant to a potential North Korean collapse (Collins 1994, 16). Following the lessons of the first Gulf War, a greater emphasis would be placed on the role of air power, particularly relying on a range of precision munitions, airborne surveillance assets, and advanced C4ISR systems, which would be better synchronized and integrated with select U.S.–ROK ground and naval components and aimed at a spectrum of strategic targets throughout North Korea.

Initially, U.S.–ROK operational plans such as the OPLAN 5027-94 envisioned halting a potential North Korean offensive at 20–30 miles south of the DMZ (FEBA Bravo) during the initial 5–15 days of hostilities, while conducting Noncombatant Evacuation Operations and force deployment of augmented units that would follow the newly initiated U.S.–ROK Reception, Staging, Onward Movement and Integration concepts. Once major reinforcements would arrive within estimated 15–40 days, the plan envisioned a U.S. Marine Expeditionary Force together with the 82nd Air Assault Division supported by ROK divisions would transfer the war toward the city of Wonsan on North Korea's east coast. A major amphibious operation by combined special U.S.–ROK forces on Wonsan would likely follow, and the force would then advance toward Pyongyang. Simultaneously, a major U.S.–ROK counteroffensive would be launched from Seoul north toward Pyongyang. Both forces would then link either in Wonsan or directly in Pyongyang, destroy the remaining KPA forces, and topple the North Korean government (Global Security 2011a). After the nuclear crisis of 1994, however, this version of the OPLAN 5027-94 was completely overhauled (OPLAN 5027-96) and subsequently updated every two years (OPLAN 5027-98; 00; 02; 04). Rather than stopping a North Korean invasion and pushing it back across the DMZ, the new concepts focused on offensive operations into North Korea with specific aim to defeat and abolish North Korea as a functioning state and "reorganize" it under South Korean control. New operational priorities also included countering chemical and biological attacks against Seoul by preemptive strikes against North Korea's military and air force bases (Global Security 2011a).

With the increasing emphasis on the RMA in the mid-1990s, the character and direction of the U.S.–ROK combined training has been also revamped. The system-of-systems concepts, for example, articulated in the *Joint Vision*

2010 (i.e., dominant maneuver, precision strike, full dimensional protection, and focused logistics) defined the underlying theme of the 1997 U.S.–ROK combined command post exercise *Ulchi-Focus Lens* (UFL '97). The UFL '97 – conceptualized as a theater simulation-based exercise involving nearly all the active Korean armed forces, the USFK, and other U.S. units that would augment Korea in the event of a crisis or conflict – integrated advanced C4I systems with select ground, maritime, air, and unconventional warfare components so that each system could exchange data and share information (Global Security 2011b). The broader systems integration achieved in the UFL '97 provided a common operating picture at select command posts throughout South Korea as well as in the United States. It also enhanced intelligence reporting, flow of information critical to battle, which led to improved situational awareness and streamlined decision making in the coordination of simulated precision strikes, focused logistics, and dominant maneuver (Global Security 2011b). According to Gen. (ret.) John H. Tilelli, then the Commander-in-Chief of the CFC (1996–1999), "in simple terms, we adopted the framework for JV 2010 experimentation using a common operating picture. The ultimate outcome sought was to maximize joint and combined relative combat power to fight and win decisively. We looked for ways to achieve synergy with combined capabilities in an asymmetric manner to offset the enemy's numerical advantage and sustain operational initiative" (1998, 77).

In the late 1990s, the accelerating RMA drive within the U.S. military continued with the development of information superiority, network-centric, and precision fire concepts. The Korean theater served as an important testing ground for applying, experimenting, and analyzing RMA-oriented activities. The various inter-service components (U.S. Army, Air Force, Navy, and Marines) experimented with new systems, technologies, and concepts in accordance with the *JV2010* – using Korea as one of the key simulation environments for modeling, wargaming, joint experimentation, and battle experiments. In October 1998, for example, the U.S. Navy 7th Fleet conducted a *Fleet Battle Experiment Delta* during the exercise *Foal Eagle* (an annual joint and combined exercise sponsored by the CFC Korea) that focused on enhancing capabilities in four key mission areas: (1) joint counterfires, (2) joint counter special operations forces, (3) amphibious operations, and (4) joint theater air defense. The experiment has been widely noted in the U.S. defense community, as it tested a new "land-sea" engagement network with new operational concepts for the counter special operations force (Alberts *et al.* 2001, 272). In particular, the U.S. Navy outfitted its forward-deployed ships with its prototype command and control application – the Land Attack Warfare System, which was networked with select U.S. ground and air force units in South Korea (Army's 2nd Infantry Division, AH-64 Apache Helicopter Squadrons from the 6th Combat Air Brigade, a range of Navy and Marine Corps units, and a Maritime Air Support Operations Center). The system simulated countering hundreds of North Korean special operations boats. In this context, the U.S. Navy attempted to translate its RMA-oriented "Ring of Fire" operational concept into the joint and combined domains – integrating intelligence information that would provide a common picture of the battlefield and monitor the status and capabilities of potential shooters, enabling the task force commander to rapidly assign the most capable weapons system from a pool of forces to engage appropriate enemy targets in a series of precision attack missions (Cohen 1998). The qualitative and quantitative results of the exercise demonstrated the potential of network-centric operations and their applicability for the Korean Peninsula. For example, the average decision cycle time was reduced from 43 to 23 minutes; average mission timeline (command and control time plus operational time) was cut in half; shooter effectiveness (kills per shot) was increased 50 percent; assets scrambled was decreased by 15 percent; and leakers (special operations vessels that passed through the engagement zone to their operational destinations) were decreased by a factor of 10 (Alberts *et al.* 2001, 273). The exercise also corrected some of the previous interoperability deficiencies, design issues, and technical problems, particularly in the integration of the various ground, air, and maritime components (Santarelli 1998, 74).

With the accelerating RMA drive in the U.S. military, which gradually permeated into the combined U.S.–ROK training, exercises, and operational requirements in late 1990s, the ROK political and military echelons began to examine the changes in the dynamics of modern warfare, the implications of the RMA, and the concomitant strategies for comprehensive defense reforms. In April 1998, the MND created the National Defense Reform Committee (NDRC) with the task to rethink nearly every aspect of the Korean defense establishment, including force structure, defense management, operational concepts, and strategic culture (Kim and Finnegan 2002, 33). The NDRC aimed to provide a comprehensive strategic blueprint to transform the ROK military into a "high-technology defense force" able to leverage select RMA technologies and maximize military effectiveness against a range of threats with efficient use of limited defense resources. Its five-year charter emphasized the following main goals: (1) creation of an advanced national defense posture, (2) improvement of military policy geared toward the North Korean threat while reducing tensions on the Korean peninsula, (3) strengthening the U.S.–ROK alliance and enhancing security cooperation with the regional countries, and (4) creating a "citizen-oriented" armed force commensurate with democratization and openness. In this context, the NDRC set up a detailed program of 58 specific reform projects, targets, and benchmarks for the MND and the ROK armed forces, which were approved by President Kim Dae Jung in June 1998 (Kim and Finnegan 2002, 36). By providing realistic benchmarks and implementation strategies, the MND hoped to translate specific objectives into viable institutional defense reforms. The NDRC would also oversee the establishment of the Defense Acquisition Office, which would consolidate all aspects of South Korea's defense acquisition and R&D processes and the creation of the Defense Digitization Program. The progress of specific programs would be reviewed annually and adjusted based on the ongoing strategic developments and reviews.

The NDRC also shaped the direction of two influential reports published by the MND in 1999: (1) *Basic Defense Policy Report: 1999–2015 (BDPR)*, a Korean version of the U.S. *Quadrennial Defense Review*, and (2) *Mid-Term Defense*

Plan: 2000–2004. The revised second version of the *BDPR*, in particular, differed from previous defense policy reports, as it reflected one of the most comprehensive efforts to date to formulate ROK's defense policy goals (Lee 2000). The *BDPR* provided long-term strategic assessment of the evolving regional security environment around the Korean Peninsula, noting that over the next two decades Northeast Asia's security will likely be defined by tensions between maritime and continental powers, as well as strategic competition between the major powers and the two Koreas. With regard to the RMA, the *BDPR* emphasized the need to transform the ROK armed forces into compact, cohesive, efficient, and high-tech force by 2015. Similarly, the *Mid-Term Defense Plan*, published in 1999 and updated in 2002, envisioned "Advanced Elite Defense" concept for the ROK forces based on the pillars of information dominance and "omnidirectional defense" adaptable to an "uncertain security environment" (ROK MND 2002).

In April 1999, the NDRC set up the influential Revolution in Military Affairs Planning Group (RMAPG) to oversee short-, mid-, and long-term defense reforms in accordance with the global and regional security dynamics and foreign military modernization trends. According to the 2000 defense white paper, "the goal of the RMA Planning Group is to create a defense strategy by analyzing the future security environment and national development trends and to research the general concept and direction of each of the following fields: battlefield management, military creed, military technology, force capability system, management system, leadership and education training" (ROK MND 2000, 134). Finnegan and Kim argue that the RMAPG, with its comprehensive, long-term focus "signaled an earnest effort to truly transform the [Korean] defense establishment" (Kim and Finnegan 2002, 37). Its three-year charter, based on a phased research, aimed at (1) conceptualizing long-term future security environment 20 to 30 years ahead, including formulating baseline assumptions of Korean unification scenarios; (2) defining the conceptual foundation for a Korean RMA; (3) providing recommendations and developing action plans for its implementation; and (4) institutionalizing the changes by updating policies and plans. The group included select officers from each ROK military service, tasked to study and benchmark service-specific RMA force modernization programs, technological capabilities, and operational concepts of advanced militaries that would be relevant for adopting in the ROK military (Hong 2005). While many of the group's recommendations and actual findings remained classified, its main ideas permeated into select MND publications, including the 2000 defense white paper, which provided a snapshot of the Korean RMA-oriented vision at that time. In particular, by analyzing the conduct and lessons learned from select conflicts – the Persian Gulf War, the Kosovo War, and the naval battle of Yangpyong – the RMAPG drew several core assumptions about the future security environment and military power: First, technology is increasingly becoming the determining factor in warfare (ROK MND 2000, 14). Combat power is derived from, and multiplied by, qualitatively advanced weapons technologies that offset quantitative advantages – remote sensor technologies, stealth technologies, long-range precision-guided munitions, integrated C4I systems coupled with NCW-oriented concepts will provide a decisive advantage in future conflicts. Second, the evolution of weapons technologies will fundamentally change the battlefield environment as well as strategy, tactics, organizations, and defense management. Specifically, the character of warfare will likely shift from traditional large-scale conventional wars or attrition warfare and resemble more dispersed, limited, or local conflicts over specific objectives. Future warfare will likely evolve into integrated information systems warfare, which will replace maneuver warfare (ROK MND 2000, 162). In order to prepare for unspecified future threats, the RMAPG recommended that the ROK military should become a "smaller but more lethal force" in three key areas: (1) building a slim-downed core force with advanced and adaptable "omnidirectional" military capabilities–in this regard, the ROK military should become a viable "combat capable force" able to meet a spectrum of military threats not only from North Korea but also a range of uncertain and ambiguous threats beyond the Korean Peninsula; (2) developing an elite, *digitized* armed forces, equipped with select high-tech weapons systems and platforms, which would be able to maximize integrated military power; and (3) a *rationalized* force with advanced defense management capacity to maximize savings and efficiency in weapons procurement, R&D, testing and evaluation, and acquisition systems and processes. Taken together, the ROK armed forces should develop comprehensive and "self-sufficient" defense capabilities, able to handle a more diverse range of missions, while having a limited regional power projection effects (Kim 2005).

Parallel to the centralized MND's RMA-oriented defense reform initiatives, the specific ROK military services – Air Force, Army, and Navy – began to formulate their long-term strategic blueprints, develop visions of the future battlefield, and conceptualize doctrines and strategies with a time span up to 2025. In a retrospective, the service-specific Korean RMA visions – a product of various working-level committees and task forces – reflected efforts to amplify the relevance of each service in line with the MND's defense reform plans. Inherently, they also projected a degree of inter-service rivalry, debate, and competition, particularly between the traditionally dominant ROKA and the ROK Air Force (ROKAF) and Navy. For example, in 1998, the ROKA published the *Army Vision 2010*, conceptualizing its RMA-oriented strategic and operational concepts in the context of countering traditional security threats as well as extended threats, including operations other than war. In the report, the ROKA proposed creating an "Integrated Force" under a single Ground Operations Command (GOC) – an idea initially supported by the NDRC – that would convert the existing command structure by merging the First and Third Armies under the GOC, creating a Rear Area Command and unified Helicopter Operations Command. However, the ROKAF and ROK Navy as well as the CFC opposed this concept, arguing that "integrated force" would effectively sustain the leading role of the ROKA in future operations and provide de facto secondary roles for the Korean Navy and the Air Force. In other words, they argued that *Army Vision 2010* does not conform to Korea's long-term defense requirements and the changing dynamics of warfare, which is likely to center on robust naval capabilities and air power. Meanwhile, the CFC objected the idea on the ground that in the absence of viable autonomous (Korean) operations plan creating the GOC would be irrelevant. It would significantly pose complications to the viability of existing war plans – that is, OPLAN 5027, which would have to be substantially modified. Amid the continuing debate in 1999, the MND eventually postponed the establishment of the GOC indefinitely (Lee 2000).

In 1999, the ROK Navy also presented its *Navy Vision 2020*, projecting the importance of network-centric warfare in the twenty-first century and its relevance for naval combat. In the same year, the ROK Joint Chiefs of Staff (JCS) published the *Joint Vision 2015*. The report focused on the need to enhance ROK's overall power projection capabilities through the integration of RMA-oriented command and control, information, maneuver, strike, logistics, and force protection. The follow-on report was the *RMA Concept and Direction of the ROK Navy* in 2001, which emphasized the importance of NCW (Hong 2005). Among the ROK military services, however, the ROKAF has arguably been the most forward-looking service in promoting the Korean RMA drive, benchmarking, adopting and adapting select U.S. information superiority, network-centric, and precision strike concepts while enhancing its air power modernization. In 1998, the ROKAF pioneered the process with the publication of the *Air Force Vision (AFV) 2030* (*Daehankongkun*), which called for attaining advanced Korean aerospace capabilities in two phases: (i) evolutionary design up to 2015 and (ii) revolutionary design up to 2025. The report projected that by 2025 the Korean peninsula will be unified and the overall defense configuration will be based on enhanced "jointness" in the context of the continuing U.S.–ROK alliance, irrespective of U.S. force deployment in Korea. The *AFV 2030* envisioned a fundamental paradigm shift in the ROK military: from a land-based force toward air power and navy-centered force; from functional force toward a mission-based force; and from a threat-based to capability-based defense planning. Subsequently, the ROKAF would have to rethink virtually all aspects of its aerospace capabilities and resource allocation for the next two decades. At the strategic level, for example, it would have to develop a new doctrine that would meet "total" force requirements while retaining the flexibility and adaptability to changes in global and regional security environment. At the operational level, the ROKAF would redesign its warfighting concepts and tactics, leveraging the combat power of select advanced weapons systems, technologies, and platforms – including space-based systems, unmanned systems, stealth technologies, and precision-guided munitions (Lee 2000).

Following the initial *AFV 2030* report, the ROKAF set up the "Air Force RMA Planning Committee" in June 1999, also referred to as the "Air Force 2025 Planning Committee." It was tasked to define key assumptions, scenarios, and requirements for the ROKAF future doctrine and strategy – in areas of battlefield management, weapons systems, organization, logistics support, people, and leadership. The committee worked closely with the ROKAF's Combat Development Group, which conceptualized the Future Air Campaign Concept (2000) and subsequently revised the *AFV 2025* (2002). The former presented a Korean emulation of U.S. NCW and IW-oriented operational concepts and future battlefield system architecture based on integrated C4I, ISR (intelligence, surveillance, and reconnaissance), combat systems, and combat support systems. The integration of various ISR collection and intelligence assets would aim at achieving "intelligence dominance" – combining early-warning, real-time long-range surveillance and intelligence processing of enemy activity while protecting friendly systems and striking enemy's information systems. The C4I component would then synchronize the various automated command and control systems, target management systems, and combat support systems and essentially translate intelligence dominance into a "decisive command and control" (i.e., prioritizing targets and coordinating precision strikes). The combination of NCW and IW concepts with relevant advanced weapons systems and technologies would provide a critical factor in the outcome of both defensive and offensive operations (Bae 2001).

The revised *AFV 2025* distilled the ROKAF's RMA vision further by providing not only a conceptual template of future warfare but also identifying key issues and areas relevant for South Korea's short- and long-term strategic planning and transformation of the ROKAF. The *AFV 2025* sharpened the baseline projections and assumptions regarding the future global and regional security environment – that is, political, economic, social, and technological drivers affecting international security over the next 25 years. Specifically, the report noted (1) the accelerating effects of globalization and decentralization of power, (2) the economic rise of Asia, (3) transition to knowledge- and information-based societies, and (4) the acceleration of IT-dominant and -driven industries. With regard to Northeast Asia, the report pointed to the qualitative changes in the military configuration of regional powers surrounding Korea (Japan, China, Russia), which are actively pursuing RMA-oriented force modernization programs aimed to enhance their regional power projection capabilities. Focusing on the emerging strategic template and military-security environment, the report also emphasized the complexity of unconventional threats, including terrorism, proliferation of WMD, ballistic missiles and stand-off weapons systems, cyber warfare, low-intensity conflicts, and new types of conflicts based on resource and energy competition. According to the *AFV 2025*, the complexity of the emerging strategic environment coupled with the ongoing regional RMA drive is likely to change the security dynamics in East Asia.

With the changing character of threats, the *AFV 2025* envisioned a concept of *extended future battlefield* with unprecedented operational tempo, lethality, and scope of the use of force. In this template, the ROKAF would be engaged in "parallel, joint, multi-dimensional operations" in five overlapping domains: (1) *land*: small-sized, dispersed specialized battles fought mainly by special forces, backed by multiple manned and unmanned platforms that utilize the best of advanced intelligence and technology; (2) *sea*: stealthy, small-scale (as opposed to large carrier group operations) on-sea and underwater operations; (3) *air*: an array of operations, including intelligence collection, reconnaissance, surveillance, target selection, combat assessment, command and control, emergency rescue, sea lines protection, anti-terrorism, long-range precision strikes using a mix of manned, unmanned, and stealth platforms launched from land and sea; (4) *space*: extending air operations and presenting new vital strategic domain, which would shape the outcome of any war or dispute at any level; (5) *cyberspace*: information operations aimed to maintain information superiority, executed in times of war and peace on a range of military and non-military targets.

In sum, the various service-specific "Korean" RMA concepts and visions of the 1990s (particularly those conceptualized by the ROKAF) represented early attempts to identify select RMA-oriented elements – concepts and technologies – that would be most relevant for the Korean strategic context in the short, mid, and long term. In the process, South Korean defense planners assessed key issues, areas, and implications of modern warfare, particularly studying the evolving U.S. RMA and adapting select concepts into their long-term defense plans and force improvement programs. The conceptual emphasis on attaining an "omni-directional, advanced force" became the cornerstone of the Korean RMA drive. While the concept did not represent a unique theoretical or conceptual innovation with regard to the RMA, it stimulated a broader national security debate on the current status and potential future direction of the ROK forces (as well as the U.S.–ROK alliance). Proponents of the Korean RMA emphasized the need to move beyond traditional threat-based defense planning in order to attain comprehensive "self-reliant" defense capabilities vis-à-vis North Korea as well as undefined future threats. Moreover, RMA proponents pointed to South Korea's outdated force structure framed around conventional ground forces (Moon and Lee 1999; Moon 2000). However, notwithstanding the increasing awareness and acknowledgment of the RMA, the ROK defense establishment continued to face structural, political, cost and budgetary constraints coupled with inter-service rivalries that precluded significant shift in the military force structure, resource allocation, training, and operational conduct. In other words, the early Korean RMA visions and concepts of the 1990s have not translated into a disruptive paradigm shift, rather they reflected a search for new security concepts both at the strategic and operational levels that would account for the increasingly changing security dynamics on the Korean Peninsula.

U.S.–ROK alliance transformation in the 2000s

In the post-9/11 era, the United States and South Korea embarked on a process of adapting the alliance to the changes in the global U.S. defense posture and reconfiguring the roles, missions, and existing command structures (Perry *et al.* 2004). At that time, the largely techno-oriented RMA conceptualization in the U.S. military morphed into a much broader process of defense transformation that aspired beyond changes in operational concepts, force structures, and equipment. The Bush administration increasingly realized that while the emergence of advanced RMA-oriented technologies may enable defense transformation processes, the progressive complexity and new set of global security challenges make transformation necessary (O'Rourke 2007). The U.S. defense transformation became a ubiquitous and comprehensive "enterprise" propelling changes in America's defense management and use of force. Strategic and operational flexibility, agility, and lethality coupled with enhanced expeditionary capabilities and development of a new generation of weapons technologies underscored the transformation imperatives.

Inherently, the direction and character of defense transformation in the early 2000s had a significant impact on U.S. allies and coalition partners worldwide, including South Korea. To begin with, U.S. defense transformation stipulated the need for realignment of the U.S. forces deployed on the Korean Peninsula, which would be gradually reconfigured toward supporting regional or even global missions rather than addressing traditional static peninsular defense (Reiss 2009). In this context, at the Korea–U.S. Security Consultative Meeting in 2002, Washington and Seoul launched the "Future of the ROK-U.S. Alliance Policy Initiative" (FOTA), aimed at devising a mutually acceptable plan to reassign existing missions and command structures. During the ensuing FOTA talks in 2004 and 2005, both the United States and South Korea agreed to transform the U.S. strategic presence and operational conduct in Korea (Nam 2007). Under the agreement, the United States would permanently remove about one-third of its existing troop strength stationed in South Korea or 12,500 troops by 2008, while the remaining 25,000 U.S. soldiers would undergo a phased relocation from the current 43 bases scattered around the country to 16 bases, concentrated in two areas south of the Han River. The USFK redeployment, including the relocation of the USFK headquarters from the Yongsan base in the center of Seoul to a new hub-base near the city of Pyeongtaek, would essentially limit the vulnerability for U.S. troops to a potential surprise attack by North Korea – that is, by lessening the "tripwire" effect of having U.S. forces close to the DMZ – but also, more importantly, provide greater latitude or "strategic flexibility" for new supporting roles, missions and capabilities that would cover wider multifunctional, expeditionary context, including regional contingencies beyond the Korean Peninsula (Przystup and Choi 2004). The unprecedented reshaping of the U.S.–ROK alliance generated significant political debates between its proponents and objectors in Korea, particularly with regard to the implications and operational effects on South Korea's deterrence – that is, some Koreans have interpreted the proposed USFK relocation and force reduction as a removal of the traditional frontline "trip-wire" deterrent and commitment to defend South Korea. Moreover, the debate created uncertainties and concerns within the South Korean military establishment on the potentially increasing technological, organizational, and conceptual interoperability gaps between the U.S. and ROK militaries. Under the revamped alliance system, the ROK forces would be required to increase their qualitative combat capabilities and assume greater autonomy and responsibility in defense of the country. For example, the ROKA would provide frontline control along the DMZ and the Joint Security Area at Panmunjom, maritime counter-infiltration operations, rapid mine-laying, search and rescue, rear-area chemical and biological decontamination, military police operations, and battlefield counter-battery artillery operations. In order to mitigate Korean concerns, the United States has offered to introduce a range of weapons technology upgrades, including the integration of advanced IRS technologies and PGM assets to South Korea (Kim 2005; Nam 2007). However, South Korea's strategic uncertainties, diverging threat perceptions, and persisting political debates not only on the future of the U.S.–ROK alliance but also regarding the future of North Korea in the early 2000s continued.

Arguably, South Korea's national security debates permeated also in the policy of "co-operative self-reliant defense" – a long anticipated goal of South Korea's defense policy dating back to early 1970s and revamped during the Roh Moo-Hyun administration. The policy emphasized the need for "the simultaneous development of the ROK-U.S. alliance," but more importantly, "a self-reliant national defense ... [when] South Korea will be equipped with capabilities and systems to play a leading role in repulsing any potential provocation" (ROK National Security Council 2004, 27). In other words, while the strategic bedrock of South Korea's security would remain anchored in the U.S.–ROK alliance, South Korea would mitigate its dependence on the United States by gradually adopting more advanced military capabilities. To implement the vision, South Korea would pursue broader military reforms, particularly in areas of command and control, force structure, and the U.S.–ROK alliance. In the long term, ROK defense planners envisioned the adoption of the concept of "sufficient defense" – having independent capability to deter any existing threats from North Korea while preparing for unspecified future threats. Underscoring the policy became the vision of transforming the ROK military into a smaller, but more responsive, networked, and digitized standing force with independent early-warning, surveillance, and reconnaissance platforms; real-time integrated C4I systems; and long-range precision strike capabilities (ROK MND 2004).

Indeed, the Mid-Term Force Improvement Plan 2003–2007, updated in April 2002, revived the concept of "Advanced Elite Defense" – a twenty-first-century combat force based on the twin pillars of "Information Dominance" and "Omni-Directional Defense," envisioned to meet a range of current as well as future-oriented uncertain threats. In contrast to the ongoing debates in South Korea's political spectrum regarding the credibility of North Korean threats mitigated by the accelerated engagement by the Roh Moo-Hyun administration toward Pyongyang, the report clearly identified North Korea as the principal military threat with "no change" in the KPA's military configuration. It noted that the North Korean military is increasing the forward deployment of its mechanized units and long-range artillery in addition to biological/chemical weapons, long-range ballistic missiles, special forces, and other asymmetrical upgrades. It also stated that the ROK is moving to an era of "uncertain security environment" marked by the "co-existence of South-North cooperation and simultaneous military confrontation." Therefore, the ROK forces should acquire and field "high performance weapon systems at a reasonable cost within the requested time frame; and obtain R&D capabilities for certain key weapon systems and thus realizing a revolution in military affairs (RMA) pertaining to defense technology" (ROK MND 2002).

In 2005, the ROK MND increased the funding for force development (South Korea's defense budget increased 13.4 percent over that of 2004). This enabled South Korea's defense planners to support a total of 181 existing programs and the commencement of 17 new programs (Johan 2004, 23). These programs emphasized four capability areas: (1) independent early-warning, surveillance, and NCW capabilities; (2) acquisition of precision-guided weapons; (3) service-specific force modernization programs; and (4) active multilayered defense capabilities to meet an array of uncertain future threats. Specifically, the service-specific programs included enhanced mobility and targeting capabilities, including the procurement of next-generation main battle tanks (MBTs), self-propelled howitzers, multi-purpose helicopters (KMH), advanced precision munitions for the ROKA; the acquisition of more advanced surface combatants such as the KDX-II and the KDX-III supported by modern air-independent propulsion submarines such as the German 214 type for the Navy; multi-role, long-range fourth-generation fighter aircraft (F-15K, F-16D) armed with stand-off PGMs and air-to-air missiles, Airborne Warning and Control System surveillance aircraft (EX), T-50 advanced trainer/light attack aircraft, and multi-layered air defense systems for the ROKAF. Underlying the process was the concept of "Defense Digitization," which stipulated a comprehensive modernization of South Korean C4ISR assets and systems. The ROK MND defined digitized defense in 2000 as "the process of transforming the overall defense structure into the information and knowledge-based network, using the latest information technologies consisting of computers and high-tech communications equipment" (ROK MND 2004, 105). *Digitized Defense*, inherently a Korean version of NCW, envisioned a creation of integrated Defense Digitization Network (DDN) that would link all tactical units, provide real-time visualization of the battlefield, real-time command and control, advanced target acquisition, and communication with sensor-to-shooter systems (Kim 2005). The DDN would be developed in three phases: (1) development of an ultra-high-speed DDN by 2005, (2) integration of joint, combined, and tactical C4ISR systems and infrastructure by 2010, and (3) comprehensive defense digitization appropriate for a knowledge-based society by 2015. The systems would rely primarily on South Korea's domestic innovation and advances in information and communications technologies.

While previous South Korean force modernization plans reflected rather an evolutionary and "largely reactive adjustments to changing security environment" (Moon and Lee 2008, 120), the subsequent Defense Reform Plan 2020 (DRP 2020) has been conceptualized in a much more ambitious scope than ever before – in terms of its aims, content, time span, and required budget (Han 2006). The original plan was published in September 2005 by the Committee on Defense Reform under former Defense Minister Yoon Kwang-ung and later modified in 2008, 2009, and 2010. The plan was strongly influenced by the French military modernization initiatives and reforms – perceived as the most relevant for the South Korean case with its comprehensive transformation from a manpower-intensive military force to a capability RMA-oriented future force. To begin with, the DRP 2020 was based on strategic assumptions that North Korean threats would inherently diminish by 2020, while potential intra-regional threats coupled with regional force modernization drives may emanate greater security challenges for South Korea. At the same time, the plan considered South Korea's declining birth rate, estimated to provide insufficient number of conscripts to sustain the current force size by 2020 (Bennet 2006). In this context, the plan projected a comprehensive strategic blueprint for transforming South Korean forces – from a medium-heavy infantry, artillery-centric, and largely static conventional army, chained to defending the DMZ, into a smaller, but more agile, professional, capability-oriented, and technology-intensive force (Pollack 2005). Specifically, the original DRP 2020 plan envisioned a gradual three-phased reduction in the quantity of South Korean military manpower by 27 percent from 690,000 to 500,000 by 2020. It proposed a reorganization of the command structure: the First ROKA and the Third ROKA would merge under a new GOC, while the Second ROKA would be transformed into a new Rear Area Operations Command. In addition, the DRP 2020 suggested the creation of a new Missile Command to address the threats posed by North Korean long-range artillery and ballistic missiles. The ROKAF and Navy would also streamline their command systems, from existing four layers of command to three. Existing ROKA infantry formations would be converted into mechanized forces with significantly enhanced mobility and firepower, improved tactical C4I, and transforming traditional division structures into more flexible division and brigade task forces with combined and joint capabilities for rapid crisis response (Karniol 2005; Bechtol 2006).

Under the DRP 2020, South Korea's armed forces would expand the scope and reach of their operational horizons. For example, the ROKA would expand the operational boundaries of a combat unit from 30 to 100 km. The ROK Navy would aim to expand its "blue-water capability" and reach even further offshore in the Pacific and in distant operating areas such as the Arabian Sea and the Indian Ocean. About one-third to one-half of existing but largely outdated major weapon systems would be effectively replaced with the next-generation weapons platforms, systems, and technologies in order to counter a wide range of threats as well as to match capabilities of regional neighbors (Dorschner 2007). Key force modernization programs would focus on the development, procurement, and integration of next-generation tanks (K-1A1, K2), multirole fighter aircraft (F15Ks), multirole helicopters (KMH), submarines, destroyer experimental vessels (KDX), surface-to-air missiles (SAM-X), early-warning systems (EX), independent precision strike assets, and the integration of digital C4ISR infrastructure (Bitzinger 2006). Many platforms as well as their components and sub-systems would stem from Korea's indigenous R&D defense industrial base, with foreign sources associated with the supply of major items and leading-edge technologies. In order to accelerate the force development, South Korea's defense budget would be increased by 11.1 percent annually through 2015, and 7.1 percent through 2020, totaling about 621 trillion won (US$ 431 billion at 2008 monetary rates) between 2006 and 2020. This would include 272 trillion won for force investment (about 40 times the 2005 force investment budget) and 349 trillion won for personnel and operations (Bennet 2006). The DRP 2020 would shape the subsequent five-year Mid-Term Defense Plans as well as the annual defense budgets, which would be adjusted based on changes in South Korea's security situation and force modernization progress (Paek 2009).

Along with a revamped force structure and advanced weapons technologies, the DRP 2020 also emphasized the need to adopt select RMA-oriented concepts in order to reach the capabilities to undertake missions currently held by the U.S. forces. As with previous "future battlefield" concepts, the ROK military closely observed recent U.S. combat experiences in Iraq and Afghanistan, which demonstrated the effects of mobile, net-centric, combined arms task forces and precision long-range fires linked to cooperative tactical targeting. With the DRP 2020, the ROK forces would adapt relevant conceptual elements and multidiscipline capabilities of the evolving American NCW. In particular, upon the completion of the reform, the South Korean military envisioned linking of an array of ISR platforms with select weapons platforms and precision munitions via advanced C4I infrastructure, thereby permitting non-linear maneuver and stand-off warfare. In doing so, the DRP 2020 and its operational concepts would remove existing gaps in the (1) combined interoperability with U.S. forces, particularly in the air power and C4ISR, and (2) joint interoperability among the three distinctly separate ROK services (Bechtol 2006).

However, with its ambitious scope, required timeline, and relatively high costs, the DRP 2020 propelled internal policy debates on the feasibility, affordability, pace, direction, character, and implementation of South Korea's defense transformation. The debate centered on the five key enduring challenges in South Korea's defense planning and management: (1) how to balance and prioritize South Korea's current operational requirements vis-à-vis North Korea with future-oriented and relatively uncertain regional threats; (2) how to ensure budgetary support and sustain projected increase in defense resource allocation required for implementing the defense reform; (3) how to streamline and reduce the ROK force structure without mitigating its capabilities; (4) how and when to transfer current wartime OPCON from the U.S. forces to South Korea; and, ultimately, (5) how to shape the future strategic template of the U.S.–ROK alliance. These questions have propelled heated political debates, amplified by unexpected developments that led to adjustments, revisions, and delays in the original DRP 2020. Specifically, the DRP 2020 underwent first two major revisions in 2008 and 2009 under the Lee Muyngbak administration, which shifted South Korea's military modernization priorities from the air and naval build-up aimed at long-term regional threats to attaining near-term capabilities to counter specific North Korean conventional and asymmetric threats, particularly focused on North Korea's WMD and ballistic missile programs. On November 24, 2008, the ROK MND presented a draft revision of the defense reform, stating that "[the ROK] military will readjust its arms acquisition and restricting schemes to properly deal with an imminent threat, namely North Korea" (Jung 2008). The decision was also influenced by the impact of the 2008 financial crisis and subsequent economic downturn, which proved baseline defense spending assumptions for DRP 2020 as unrealistic. Indeed, the original plan called for defense spending to increase by 9.9 percent from 2006–10, by 7.8 percent from 2011–15, and by 1 percent from 2016–20. However, spending growth has not reached this planned level – with increases of 6.98 percent in 2006, 9.03 percent in 2007, 4.16 percent in 2008, and 8 percent in 2009 (Grevatt 2008).

The revised draft downsized select procurement programs such as the K2 MBTs, readjusted the timeline and size of troop reductions, placed an emphasis on more gradual defense spending, and focused on North Korea's nuclear and missile threats. These modifications were formalized in the 2009 official revision, presented on

June 26 by Defense Minister Lee Sang-hee. Under the modified plan, South Korean military would develop capabilities in the areas of (1) surveillance and reconnaissance, (2) precision strike, (3) interception, and (4) force protection, primarily to enable preemptive precision strike capabilities vis-à-vis North Korean nuclear and missiles sites. In other words, South Korea would focus on advanced early-warning systems to detect imminent ballistic missile attacks, joint air-to-ground precision munitions, and anti-missile defense systems (KBS 2009).

In 2010, however, two unprecedented incidents have further reshaped the character and direction of South Korea's DRP 2020 military modernization trajectory. The first was the deliberate sinking of the 1,200-ton South Korean corvette *CheonAn* in the Yellow Sea near the disputed Northern Limit Line on March 26 by a torpedo attack originating from a North Korean submarine. The incident, which killed 46 sailors, raised questions about South Korea's combat readiness and responsiveness, particularly in its naval capabilities, anti-submarine warfare, command, control, and communications. The attack caught the ROK Navy by a total surprise – as its anti-submarine detection system, including sonar and early-warning systems, coupled with existing operational concepts proved ineffective in spotting the North Korean submarine operating in shallow waters. More importantly, the ROK Joint Chiefs of Staff were criticized for their relatively slow and uncoordinated response to the attack. The lack of combat readiness amplified by operational deficiencies in the ROK military became apparent also in the second major incident of 2010 – the North Korean coordinated dual artillery/rocket attack on South Korea's Yeonpyeong Island on November 23, 2010. North Korea fired about 170 artillery shells toward the island in one of the most serious confrontations since the end of the Korean War. While the ROKA units on the Island returned fire with about 80 shells, these were directed at North's coastal artillery bases at Mudo, targeting command posts and barracks, not at the actual site of mobile rocket launchers (The Chosun Ilbo 2010).

Proactive deterrence debate in 2010s

Both crises had a significant impact on South Korea's defense planning. On May 4, 2010, President Lee Myung-bak announced a military-wide review of ROK's defense posture (Falletti 2010). The 2010 National Security Review provided 71 recommendations to improve South Korea's early-warning and C4ISR capabilities, missile defenses, quality of reservists training, force integration, and command and control. More importantly, the impact of both crises eventually prompted South Korea to request a postponement of the planned transfer of OPCON to Seoul until December 2015. The decision to delay the OPCON transfer also shaped the subsequent "Strategic Alliance 2015" (SA2015) base plan, a new five-year U.S.–ROK roadmap that would "enable better synchronization of the alliance transformation efforts" (Sharp 2010). The SA2015 emphasized the need to address existing operational deficiencies, update warfighting concepts, adjust weapons procurement and training to maximize the operational effectiveness of the U.S.–ROK alliance and prepare for a wider range of contingencies. In other words, the SA2015 was designed to better prepare the South Korean military for the OPCON transition in 2015. Subsequently, in March 2011, South Korea's Ministry of Defense publically announced a new force modernization plan – Defense Reform 307, primarily addressing medium- to long-term ROK military readiness to counter potential North Korean asymmetric provocations, infiltrations, and attacks similar to the sinking of the *CheonAn* and artillery attack on the Yeonpyeong Island. The plan's strategic concept emphasized "proactive deterrence," aimed at deterring future North Korean provocations, partly by avoiding serious damage from potential provocations and by having effective retaliatory capabilities for such provocations. At the time of this writing, Defense Reform 307 has been under review under President Park Geun-hye administration and will likely be adjusted based on the security developments in and around the Korean Peninsula.

Indeed, given its continuing political and socioeconomic isolation, North Korea's military is continuously adapting its strategies and capabilities toward forms of asymmetric negation, probing any vulnerability in the U.S.–ROK alliance in order to counter its qualitatively superior technological advantages. In addition to nuclear and ballistic missile programs, North Korea has been developing cyber-related offensive military capabilities. Computer network cyber operations, both offensive and defensive, coupled with information warfare strategies and tactics provide new types of "force multipliers" for North Korea and can be viewed as new "weapons of mass effectiveness." In 2010s, North Korea has pursued three types of cyber operations vis-à-vis South Korea: (1) cyber espionage to obtain information and intelligence about U.S.–ROK military technologies, capabilities, and strategies; (2) computer network attacks aimed at denying, disrupting, or destroying South Korea's information infrastructure; and (3) information and deception operations to shape broader internal and external strategic communications, perceptions, and narratives. In the cyber espionage category, North Korea's primary overseas intelligence gathering unit, operating under the State Security Agency, is believed to increasingly rely on cyber-related techniques for cyber espionage to access information, steal sensitive data, and monitor foreign communications. Among its elite military cyber units, North Korea's cyber espionage is led by the hacker Unit 121, operating under the North Korean Army General Staff's Reconnaissance Bureau. In 2009, South Korean National Intelligence Service and the Defense Security Command reported that Unit 121 intercepted confidential defense strategy plans, including OPLAN 5027 detailing U.S.–ROK responses to potential North Korean provocations. In the same year, North Korean hackers reportedly stole information from the South Korean Chemical Accidents Response Information System (CARIS) developed by the National Institute of Environmental Research under the Ministry of Environment after infiltrating the ROK Third Army headquarters' computer network and using a password to access CARIS' Center for Chemical Safety Management.

In the category of computer network attacks, North Korea has attempted to disrupt South Korea's sophisticated digital information infrastructure using cyber-attacks to shut down major websites, disrupt online services of major banks, and probe

South Korea's readiness to mitigate cyber-attacks. These include cases of distributed denial-of-service (DDoS) attacks against four-dozen targets in South Korea and the United States in 2009 as well as "Ten Days of Rain" DDoS attacks targeting South Korean government websites and networks of the USFK lasting 10 days in 2011. The combination of clearly defined targets, highly destructive malware code, multiple encryption algorithms, and multi-tiered botnet architecture preconfigured for specific duration has led to a conclusion that the attack was set up by North Korea to test and observe how rapidly the attack would be discovered, reverse-engineered, and mitigated. Last but not least, in the category of information warfare operations, North Korea has relied on classical deception to alter the perceptions of its strategic plans. Prior to its rocket launch in December 2012, and subsequent third nuclear test in February 2013, North Korea manipulated news stories as part of a deliberate deception campaign to hide its real intentions. In case of the 2012 rocket launch, North Korea manipulated the timing of the launch so that U.S. intelligence satellites would not be overhead. Pyongyang announced technical problems with the rocket several days before the launch. At that time, U.S. spy satellites observed the North Koreans taking apart the three-stage rocket and moving the parts away from the launch pad. North Korea, however, launched the rocket without any delay, catching U.S.–ROK military and intelligence agencies off-guard.

In conclusion, for nearly seven decades, U.S.–ROK strategy on the Korean Peninsula has remained relatively constant, maintaining robust forward-active presence coupled with bilateral alliances to ensure peace, stability, and prosperity in the region. Since the end of the Cold War, however, East Asia's regional strategic template has become progressively more complex and multifaceted with the confluence of unresolved historical legacies in traditional flashpoints such as the Korean Peninsula, Taiwan Straits, territorial disputes in East China and South China seas, as well a range of nontraditional security challenges such as energy and cyber security. Notwithstanding North Korea's ongoing strategic and operational challenges, however, it is also China's increasing power projection capabilities embedded in the PLA's growing technological developments, including long-range precision strike assets, that is gradually redefining regional military balance and subsequently U.S.–ROK strategy and military modernization. In particular, China's asymmetric "counter-intervention" concepts and weapons technologies, designed to deny U.S. forces and its allies the freedom of action in China's "near seas" by restricting deployments of U.S. forces into theater (anti-access) and denying the freedom of movement of U.S. forces already there (area denial), amplify the magnitude of strategic and operational challenges for U.S.–ROK forces in the region as they become increasingly vulnerable to PLA's long-range precision strike capabilities. Moreover, as North Korea develops offensive cyber warfare capabilities, future conflicts on the Korean Peninsula will be linked with confrontations in and out of cyberspace, cyber-attacks on physical systems, and processes controlling critical information infrastructure, information operations, and various forms of cyber espionage. Deterring these "hybrid" threats, whether nuclear or cyber-related, will be increasingly challenging. While South Korea has established a new cyber warfare command designed to counter North Korean cyber threats, South Korea has been searching for a comprehensive strategy and military innovation, aiming to engage instruments of both soft power of diplomacy and hard power of military force to shape conditions for change in and around the Korean Peninsula.

Note

1 This chapter is partly based on Raska, Michael, "RMA Paths and Patterns in South Korea's Military Modernization," *Korean Journal of Defense Analysis* 23, no. 3 (2011): 369–85. Republished with permission of the publishers.

References

Alberts, David, John Garstka, Richard Hayes, and David Signori. 2001. *Understanding Information Age Warfare*. Washington, DC: DoD C4ISR Cooperative Research Program.

Bae, Chang Sik. 2001. "The Direction of Development of Airborne Early Warning System Required on the Korean Peninsula." Paper presented at the 4th International Conference on Korean Air Power: Strategic Intelligence and Air Power Superiority, August 30, Yonsei University, Seoul.

Bae, Jong-Yun, and Chung-in Moon. 2005. "Unraveling the Northeast Asian Regional Security Complex: Old Patterns." *Korean Journal of Defense Analysis* 17 (2): 7–34.

Bechtol, Bruce. 2006. "Force Restructuring in the ROK-US Military Alliance: Challenges and Implications." *International Journal of Korean Studies* X (2): 19–41.

Bennet, Bruce. 1993. *Global 92 Analysis of Prospective Conflicts in Korea in the Next Ten Years*. Santa Monica, CA: RAND.

——. 2006. *A Brief Analysis of the Republic of Korea's Defense Reform Plan*. Occasional Paper 165, 1–55. Santa Monica, CA: RAND.

Bitzinger, Richard. 2006. *Transforming the U.S. Military: Implications for the Asia-Pacific*. Barton, Australia: Australian Strategic Policy Institute.

Blackwill, Robert, and Paul Dibb. 2000. *America's Asian Alliances*. Cambridge, MA: MIT Press.

Boulding, Kenneth. 1978. *Stable Peace*. Austin, TX: University of Texas.

Cha, Young-Koo. 1994. "National Security Strategy of South Korea: Looking Toward the 21st Century." In *Asia in the 21st Century: Evolving Strategic Priorities*, edited by Michael Bellows, 71–91. Washington, DC: National Defense University Press.

Cho, Myung Hyun. 1989. *Korea and the Major Powers: An Analysis of Power Structures in East Asia*. Seoul, Korea: Research Center for Peace and Unification of Korea.

Clough, Ralph. 1976. *Deterrence and Defense in Korea*. Washington, DC: Brookings Institution.

Cohen, William. 1998. *SecDef Annual Report: Annual Report to the President and the Congress – 1999*. Washington, DC: Office of the Secretary of Defense.

Collins, John. 1994. "Korean Crisis, 1994: Military Geography, Military Balance, Military Options." *CRS Report for Congress* (94-311S): 1–22.

Combined Forces Command. 1995. *Deep Operations Primer - Korea*. Seoul, Korea: CFC Korea, Operations Division.

——. 2002. *Air-Ground Operations Korea (CFC PUB 3-2.2)*. Seoul, Korea: CFC Headquarters.

Curtis, Gerald, and Sung-Joo Han. 1983. *The U.S.-South Korean Alliance: Evolving Patterns in Security Relations*. New York, NY: Lexington Books.

D'Amico, Robert. 1999. "Joint Fires Coordination: Service Competencies and Boundary Challenges." *Joint Forces Quarterly* 21 (Spring): 70–77.

Dorschner, Jim. 2007. "South Korea: Widening Horizons." *Jane's Defense Weekly*, June 20.

Erickson, Andrew. 2012. "China's Modernization of Its Naval and Air Power Capabilities." In *Strategic Asia 2012–13: China's Military Challenge*, edited by Ashley Tellis and Tanner Travis, 60–125. Washington, DC: The National Bureau of Asian Research.

Falletti, Sebastien. 2010. "Country Briefing: Seoul Searching." *Jane's Defense Weekly*, July 27.

Glaser, Bonnie, Scott Snyder, and John Park. 2008. *Keeping an Eye on an Unruly Neighbor: Chinese Views of Economic Reform and Stability in North Korea*. Washington, DC: Center for Strategic and International Studies.

Global Security. 2011a. "OPLAN 5027 Major Theater War – West". Military. Accessed July 1, 2014. http://www.globalsecurity.org/military/ops/oplan-5027.htm.

———. 2011b. "Ulchi-Focus Lens". Military. Accessed July 1, 2014. http://www.globalsecurity.org/military/ops/ulchi-focus-lens.htm.

Godement, Francois. 2005. "North-East Asia: Time to Rethink?" *Disarmament Forum* (2): 5–11.

Grevatt, Jon. 2008. "South Korea's 2020 Spending Plan is Under Review." *Jane's Defense Industry*, December 1.

Han, Yong Sup. 1999. "South Korea's Strategy of Cofnlict: The Past, Present, and Future." In *Strength Through Cooperation: Military Forces in the Asia-Pacific Region*, edited by Frances Omori and Mary Sommerville. Washington, DC: National Defense University Press.

———. 2006. "Analyzing South Korea's Defense Reform 2020." *Korean Journal of Defense Analysis* 18 (1): 111–34.

Hodge, Homer. 2003. "North Korea's Military Strategy." *Parameters* 33 (1): 68–81.

Hollis, Patrecia. 1993. "The Korean Theater – One-of-a-Kind: Interview With General Robert W. RisCassi." *Field Artillery*, February, 7–11.

Hong, Sung-Pyo. 2005. "Impact of Information Technology Revolution on Revolution in Military Affairs in the ROK Armed Forces." In *Bytes and Bullets: Information Technology Revolution and National Security on the Korean Peninsula*, edited by Alexandre Mansourov, 218–33. Honolulu, HI: Asia Pacific Center for Security Studies.

IISS. 2010. *The Military Balance 2010*. London, UK: International Institute for Strategic Studies.

Johan, Saad. 2004. "ROK Armed Forces: Force Development Continues Unabated." *Asian Defence Journal* (10): 22–26.

Johnson, Stuart, and Joseph Yager. 1979. *The Military Equation in Northeast Asia*. Washington, DC: Brookings Institution.

Jung, Sung-Ki. 2008. "South Korea to Overhaul Modernization Plan." *Defense News*, December 15.

Karniol, Ron. 2005. "South Korea Revives Radical Force Restructuring Strategy." *Jane's Defense Weekly*, October 5.

KBS. 2009. "Defense Reform 2020 Plan Revised." *KBS World*, July 7.

Kim, Jae-Chang. 2003. "The New International Order and the U.S.–ROK Alliance." *Korean Journal of Defense Analysis* 15 (2): 57–75.

———. 2005. "Digital Defense: Goals, Missions, Achievements." In *Bytes and Bullets: Information Technology Revolution and National Security on the Korean Peninsula*, edited by Alexandre Mansourov, 136–42. Honolulu, HI: Asia Pacific Center for Security Studies.

Kim, Jae Chang, Interview by Michael Raska. 2007. Interview with ROKA Gen. (ret.) JC Kim. June 8.

Kim, Jiyul, and Michael Finnegan. 2002. "The Republic of Korea Approaches the Future." *Joint Forces Quarterly* (Spring): 33–40.

Kim, Koo Sub. 2009. "Substance of North Korea's Military Threats and the Security Environment in Northeast Asia." *Korean Journal of Defense Analysis* 21 (3): 239–50.

——. 2010. *"Substance of North Korea's Military Threats and the Security Environment in Northeast Asia." Korean Journal of Defense Analysis 21 (3): 239–50.*

Kim, Mikyoung. 2005. "The U.S. Military Transformation and Its Implications for U.S.–R.O.K. Alliance." *IFANS Review* 13 (1): 15–39.

Kim, Min Seok. 2010. "North Adopts New War Invasion Strategy." *Joonang Daily*, April 27.

Kim, Samuel. 2006. "China's Conflict-Management Approach to the Nuclear Standoff on the Korean Peninsula." *Asian Perspective* 30 (1): 5–38.

Klinger, Bruce. 2008. "Evolving Military Responsibilities in the U.S.–ROK Alliance." *International Journal of Korean Studies* 7 (1): 25–42.

Lee, Chung Min. 1988. "Holding the 'Hollingsworth Line': Conventional Deterrence and Defense in the Korean Peninsula." In *The U.S.-Korean Security Relationship: Prospects and Challenges for the 1990s*, edited by Harold Hinton, Donald Zagoria, Jung Ha Lee, Gottfried-Karl Kindermann, Chung Min Lee, and Robert Pfaltzgraff. New York, NY: Pergamon-Brassey's for the Institute for Foreign Policy Analysis.

——. 2000. "Future of ROK-U.S. Security Relations and Command Structure Under Conditions of Mutual Coexistence on the Korean Peninsula." Paper presented at the Korea-U.S. International Conference on "Practical Steps from War to Peace on the Korean Peninsula", New York, NY, September 20.

——. 2001. "North Korean Missiles: Strategic Implications and Policy Responses." *Pacific Review* 14 (1): 85–120.

——. 2005a. "In Search of Strategy: South Korea's Struggle for New Security Paradigm." *Disarmament Forum* (2): 13–23.

——. 2005b. "The North Korean Missile Threat and Missile Defense in the Context of South Korea's Changing National Security Debate." *Comparative Strategy* 24 (3): 253–75.

Lee, Chung Min, and Jonathan Pollack. 1999. *Preparing for Korean Unification: Scenarios & Implications*. Santa Monica, CA: RAND.

Lee, Manwoo, Ronald McLaurin, and Chung-in Moon. 1988. *Alliance Under Tension: The Evolution of South Korean-U.S. Relations*. Boulder, CO: Westview Press.

Luttwak, Edward, and Steven Canby. 1983. "The Defense of Korea." In *Northeast Asia in the 1980s: Challenge and Opportunity*, edited by Robert Downen, 23–25. Washington, DC: Center for Strategic and International Studies.

McLaurin, Robert, and Chung-in Moon. 1989. *The United States and the Defense of the Pacific*. Boulder, CO: Westview Press.

Minnich, James. 2005. *The North Korean People's Army: Origins and Current Tactics*. Annapolis, MD: Naval Institute Press.

Minnick, Wendell. 2014. "China Mixing Military Modernization With Tailored Coercion." *Defense News*, April 14, 1.

Moon, Chung-in. 1998. "South Korea: Recasting Security Paradigms." In *Asian Security Practice: Material and Ideational Influences*, edited by Muthiah Alagappa, 264–87. Stanford, CA: Stanford University Press.

——. 2000. "Changing Threat Environment, Force Structure, and Defense Planning." In *Emerging Threats, Force Structure, and the Role of Air Power in Korea*, edited by Natalie Crawford and Chung-in Moon, 89–114. Santa Monica, CA: RAND.

Moon, Chung-in, and Chung Min Lee. 1999. *Air Power Dynamics and Korean Security.* Seoul, Korea: Yonsei University Press.

Moon, Chung-in, and Jin-Young Lee. 2008. "The Revolution in Military Affairs and the Defense Industry in South Korea." *Security Challenges* 4 (4): 117–34.

Nam, Chang-Hee. 2007. "Realigning the U.S. Forces and South Korea's Defense Reform 2020." *Korean Journal of Defense Analysis* 19 (1): 165–89.

Oh, Kongdan, and Ralph Hassig. 1999. "North Korea Between Collapse and Reform." *Asian Survey* 39 (2): 287–309.

Olsen, Edward. 1988. *U.S. Policy and the Two Koreas.* Boulder, CO: Westview Press.

O'Rourke, Ronald. 2007. *Defense Transformation: Background and Oversight Issues for Congress.* Washington, DC: Congressional Research Service Report Series.

Paek, Jae. 2009. "Defense Budget Layout for Successful Promotion of Defense Reform." *KIDA ROK Angle*, November 2.

Perry, Charles, Jacquelyn Davis, James Schoff, and Toshi Yoshihara. 2004. *Alliance Diversification & the Future of U.S.-Korean Security Relationship.* Cambridge, MA: Institute for Foreign Policy Analysis.

Pitt, David. 1987. "Seoul, U.S. Forces, and the North: The Balance is Delicate as Ever." *The New York Times*, April 8.

Pollack, Jonathan. 2005. "The Strategic Futures and Military Capabilities of the Two Koreas." In *Strategic Asia 2005–06: Military Modernization in an Era of Uncertainty*, edited by Ashley Tellis and Michael Wills, 137–74. Washington, DC: The National Bureau of Asian Research.

Pollack, Jonathan, and Young-Koo Cha. 1994. *A New Alliance for the Next Century: The Future of U.S.-Korean Security Cooperation.* Santa Monica, CA: RAND.

Przystup, James, and Kang Choi. 2004. "The U.S.–ROK Alliance: Building a Mature Partnership." *Institute for National Strategic Studies Special Report* (March Issue): 1–8.

Pullan, Richard. 2005. "How to Rethink War: Conceptual Innovation and AirLand Battle Doctrine." *Journal of Strategic Studies* 28 (4): 679–702.

Reiss, Mitchell. 2009. "Drifting Apart? The U.S.–ROK Alliance at Risk." *Korean Journal of Defense Analysis* 21 (1): 11–31.

Roehrig, Terence. 2006. *From Deterrence to Engagement: The U.S. Defense Commitment to South Korea.* Lanham, MD: Lexington Books.

ROK Ministry of National Defense (MND). 1998. *Defense White Paper.* Seoul, Korea: Ministry of National Defense.

——. 1995. *Defense White Paper 1994–1995.* Seoul, Korea: Ministry of National Defense.

——. 1996. *Defense White Paper 1995–1996.* Seoul, Korea: Ministry of National Defense.

——. 2000. *Defense White Paper 2000.* Seoul, Korea: Ministry of National Defense.

——. 2002. *2003–2007 Mid-Term Defense Plan.* Seoul, Korea: Ministry of National Defense.

——. 2004. *Defense White Paper 2004.* Seoul, Korea: Ministry of National Defense.

ROK National Security Council. 2004. *Peace, Prosperity and National Security: National Security Strategy of the Republic of Korea.* Seoul, Korea: Ministry of National Defense.

Santarelli, Eugene. 1998. "Air Component Support to Joint Exercises." *Joint Forces Quarterly* (Autumn/Winter Issue): 71–75.

Scalapino, Robert. 2010. "Relations Among Asian Nations and the Role of Frontiers."

Korean Journal of Defense Analysis 21 (2): 123–36.

Schreer, Benjamin. 2013. *Planning the Unthinkable War: 'AirSea Battle' and Its Implications for Australia.* Barton, Australia: Australian Strategic Policy Institute.

Scobell, Andrew, and John Sanford. 2007. *North Korea's Military Threat: Pyongyang's Conventional Forces, Weapons of Mass Destruction, and Ballistic Missiles.* Carlisle, PA: Strategic Studies Institute.

Sharp, Walter. 2010. "U.S. Forces Korea: Transforming to Meet New Challenges." *Army*, October 1.

Simon, Sheldon. 1993. *East Asian Security in the Post-Cold War Era.* Armonk, NY: M.E. Sharpe.

Snyder, Scott, and Byun See-Won. 2013. "Seeking Alignment on North Korean Policy." *Comparative Connections* 15 (1): 101–10.

Stratfor. 2010. "North Korea: Managing the Aftermath of the Cheonan Incident." *Stratfor Global Intelligence,* May 24.

Suh, Jae-Jung. 1999. "Blitzkrieg or Sitzkrieg? Assessing a Second Korean War." *Pacifica Review: Peace, Security & Global Change* 11 (2): 151–76.

Tellis, Ashley. 2012. "Uphill Challenges: China's Military Modernization and Asian Security." In *Strategic Asia 2012–13: China's Military Challenge*, edited by Ashley Tellis and Tanner Travis, 3–24. Washington, DC: The National Bureu of Asian Research.

The Chosun Ilbo. 2010. "Military Knew of North Korean Artillery Move Before Attack." *Chosun Ilbo*, November 26.

Tilelli, John. 1998. "Putting JV 2010 Into Practice." *Joint Forces Quarterly* (Autumn/Winter Issue): 76–80.

U.S. Department of Defense. 2000. *Report to Congress: Military Situation on the Korean Peninsula.* Washington, DC: Office of the Secretary of Defense.

Yun, Sun. 2013. "March West: China's Response to the U.S. Rebalancing." *Brookings Upfront*, January 31.

5 A structured-phased evolution

The 3G+ force transformation of the SAF

In the five decades since its independence as a small city-state on August 9, 1965, Singapore has grappled with insecurity and geostrategic uncertainty shaped by the confluence of historical legacies, changes in the regional and global security environment, and concomitant internal political, socioeconomic, and military imperatives. On one hand, while Singapore has experienced considerable limitations in balancing its security needs with policies directed at maintaining economic growth, the city-state has, on the other hand, gradually developed considerable economic, technological, and military resources that have amplified Singapore's ability to align its interests with great powers, sustain a credible defense posture, and strengthen its sovereignty. In this context, Singapore has devised a robust defense posture based on twin pillars of deterrence and diplomacy, which have evolved with Singapore's threat perceptions in both conventional and nontraditional domains. The key objective of this policy has been to protect Singapore's sovereignty and territorial integrity while ensuring peace and stability in the region (Tan 2002). Historically, the SAF has played an important role in both dimensions – actively developing and maintaining defense diplomacy with other armed forces while developing select capabilities that serve as a deterrent to potential threats. At the same time, Singapore's defense management capacity when it comes to planning, organizing, leading, and controlling armed forces and their supporting systems has been equally important. The SAF has sought to build a technologically advanced, operationally ready force capable of responding to a wide range of security threats (Lim 2004). Under the conceptual umbrella of the "3G fighting force" or 3G military transformation that was officially announced in 2004, the SAF has gradually pursued an RMA-oriented military modernization aimed at leveraging military-technological interoperability, adaptability, and overall combat power of its army, navy, and air force units. Indeed, driven by the increasing convergence of conventional, asymmetric, and nontraditional security challenges, reflected in ongoing geopolitical tensions among great powers, regional territorial disputes in the South and East China seas, and subsequent rising defense expenditures and diffusion of power projection capabilities in the region, the SAF's operational mission templates and training requirements have increased along with concomitant resource allocation needs. In the decade of developing 3G concepts, organizations, and technologies, the SAF has placed a greater emphasis on "spiral development" of its "force multiplier" capabilities applicable in areas such as C4ISR and battlefield management, intelligence, precision strike, unmanned systems, as well as cyber warfare. In the process, however, the SAF has inherently faced a number of challenges: at the strategic level, the question of how to strike a balance between preserving tried-and-tested strategies and force structures (based on a traditional conscription model) with implementing advanced operational concepts, organizational structures, and technologies relevant for the type of future "hybrid conflicts." At the operational level, the SAF has demonstrated greater technological dependencies that have, on one hand, propelled excellent "joint" tactical experimentation but, at the same time, invoked growing perceptions of "technological superiority" as the primary determinant, enabler, and catalyst of military effectiveness.

Based on this introductory perspective, this chapter attempts to project (1) the essential contours of the continuity and change in Singapore's security paradigm and defense strategy, (2) the intellectual history and processes that have shaped the 3G SAF trajectory, and (3) concepts of future warfare presently emerging in Singapore's strategic thought. In doing so, the chapter's principal argument is two-fold: First, the character of Singapore's force modernization trajectory has projected a continuous evolutionary path in terms of its systems and structures – a structured-phased approach: the 1G SAF (1960s–1970s) focused on capability development of individual services; the 2G SAFG reflected a period of consolidation and adaptation from service-oriented strategic thinking toward conventionally oriented combined arms warfare (1980s–1990s); and the 3G SAF (2000s onwards) has aimed at implementing a transition toward a joint strategy for multi-mission type forces with capabilities ranging from defense diplomacy to select kinetic integrated strike capabilities against a wide spectrum of threats. In the process, SAF's doctrinal orientation and operational conduct has shifted significantly in its character: from a purely island defensive "poisoned shrimp" strategy in the 1970s, which envisioned high-intensity urban combat to impose unacceptable human and material costs to potential aggressors, toward "porcupine" strategy of the 1980s that developed a limited power projection in Singapore's near-seas and envisioned a preemptive posture by transferring a potential conflict into enemy's territory, to the ongoing 3G concepts in the 2010s analogous to a "dolphin" strategy – a "smart" or networked SAF leveraging not only precision fires, maneuver, and information superiority capabilities but also defense diplomacy in diverse military operations in geographically distant areas from Singapore. However, notwithstanding the continuous experimentation, increasing technological sophistication, and participation by the SAF in overseas operations other than war – the absence of modern (or any) real combat experiences, the lack of inter-service debates, and general adherence to the established norms and strategic culture that promotes hierarchy and discipline – has arguably precluded the implementation of a more disruptive doctrinal and organizational innovation that would reflect a greater diversity in strategic thinking. In searching for military innovation, the SAF has attempted to balance between preserving tried-and-tested strategies and structures with adapting select innovative operational concepts and organizational structures based on RMA-oriented military-technological modernization in preparation for multilevel conflicts. The key dilemma facing Singaporean defense planners has been the question how to build a force and doctrine capable of dealing simultaneously with current security threats while anticipating future challenges. That said, the SAF's gradual conceptual, organizational, technological, and operational 3G force transformation and its resulting capabilities have continued to qualitatively outpace its neighbors in relative terms – its armed forces remain the best trained, best equipped, and likely the most effective in Southeast Asia.

Second, the trajectory of military innovation in Singapore transcends purely military-technological and operational domains – it must be situated in the broader context of strategic nexus of Singapore's external foreign/defense and internal public policies that shaped Singapore's readiness and responses to the gradually changing global and regional security challenges. The principal strategy underlying Singapore's security has been the concept of "Total Defense" (TD), a form of national security strategy that has for over 30 years aimed at strengthening and mobilizing resources in five mutually supportive defense domains: military, civil, economic, social, and psychological. In this context, Singapore has been able to offset an array of potential security predicaments through a *comprehensive defense framework* embedded in civil–military strategic interactions at various levels. These include the symbiotic relationships, links, and interdependencies between the various players in Singapore's "defense ecosystem," that is, the users (SAF), developers (i.e., MINDEF, DSTA, DSO dual-use R&D labs), and producers (local defense industries – that is, ST Engineering). At the same time, "TD" has aimed on building a long-term civil–military resilience through the integration of police and emergency services, civil management and oversight agencies, judicial and law enforcement institutions, fire and ambulatory services into "civil defense force," augmented by Singapore's educational system, civil service, media information sphere, and the entire civilian population. However, with the increasing regional tensions and strategic uncertainties coupled with gradual yet profound changes in Singapore's internal political and socioeconomic fabric over the past five years, the effectiveness of Singapore's "TD" strategies is increasingly challenged, particularly in terms of managing and responding to potentially more severe, cascading, multilevel crises – whether internal or external.

The contours of Singapore's traditional security paradigm

Apart from the publication of two official documents, *Defense of Singapore 1994–95* (Ministry of Defense Singapore 1995) and *Defending Singapore in the 21st Century* (Ministry of Defense Singapore 2000), coupled with occasional commemorative books and monographs published by the SAF Air Force, Navy, and Army (Republic of Singapore Navy 2007, 2008), there has been no other *official* document to date that would project a comprehensive account outlining Singapore's security and defense strategy, force structure, combat doctrine, and operational conduct (Koh Swee Lean 2014). However, it is still possible to gain insights into Singapore's evolving strategic thought through the reflections, public statements, speeches, interviews, and press releases of its leading policy makers over the years. Indeed, the writings and speeches of Lee Kuan Yew, Goh Keng Swee, S. Rajaratnam – as principal architects of the modern-day Singapore – shed light in the formulation of Singapore's foreign and defense policies from the early years of independence. More importantly, their strategic thought has arguably shaped the next generation of Singapore's political and defense leaders as reflected in select public speeches, statements, and interviews of Singapore's current and former defense ministers, SAF's Chiefs of Staff, and Service Chiefs. Coupled with academic journal articles and secondary sources assessing Singapore's defense policy, armed forces, defense industry, and trends in regional security, it is possible to contour the principal conditioning factors, evolving security challenges, and threat perceptions that have shaped the continuity and change in Singapore's security paradigm.

Geostrategic conditioning factors

In a broad perspective, Singapore's traditional security paradigm is based on the confluence of both internal and external conditioning factors, embedded in the historical legacies, geostrategic and demographic constraints, and considerable resource dependencies that have shaped Singapore's threat perceptions (Huxley 2000; Ng 2005). The common thread in both external and internal aspects is the *historical experience of vulnerability* – nearly an existential "angst" narrative, originating from the traumatic experiences of the fall of Singapore under the British colonial rule in 1942 and subsequent Japanese occupation and wartime deprivation until 1945, to the experiences in attaining independent international status in 1965, followed by challenges in creating a robust defense capability and socioeconomic stability in the immediate postindependence period. In this context, Singapore's independence has been portrayed in the literature as accidental, unwanted, and unexpected (Lee 2008). S. Rajaratnam, Singapore's first foreign minister, stated that an independent Singapore had a "near-zero chance of survival – politically, economically, or militarily," while Lee Kuan Yew himself called the idea of an independent and separate Singapore "a political, economic and geographical absurdity" (Tan 2001, 2).

A detailed analysis of the events surrounding Singapore's independence would go beyond the scope of this chapter. However, one could argue that the contesting visions, political and socioeconomic differences, racial tensions, and escalating strategic divergence between Singapore and the central government of the Federation of Malaysia in Kuala Lumpur, which gained independence from the British in 1957, coupled with threats of Communism and upheavals of Indonesia's confrontation agenda during the early 1960s led to a negotiated separation of Singapore from Malaysia on August 9, 1965. Prior to that, Singapore achieved self-governance in 1959, when Lee Kuan Yew was elected Prime Minister; it has remained a British colony until September 1963, when Singapore merged with the Federation of Malaysia. Perhaps more importantly, in Lee Kuan Yew's perspective, the consequences of Singapore's independence in 1965 represented not only an end of a vision of united and geopolitically secure

Malaya but a realization of an acute sense of strategic uncertainty and vulnerability. As Lee explained in his memoirs:

> All of a sudden, on 9 August 1965, we were out on our own as an independent nation. We had been asked to leave Malaysia and go our own way with no signposts to our next destination. We faced tremendous odds with an improbable chance of survival. Singapore was not a natural country but man-made, a trading post the British had developed into a nodal point in their world-wide maritime empire. We inherited the island without its hinterland, a heart without a body.
>
> (2000, 19)

It is important to note, however, that in the aftermath of its independence "Singapore was hardly defenseless against external threats, given not only the close defense relationship with Malaysia, but also the massive British military presence" (Huxley 2000, 8). The British have built Singapore's original military infrastructure – a naval base and coastal defenses during 1920s and 1930s to protect the island. Since 1927, Singapore has owned a system of defense that included the army, navy, and air force. From 1948–1960, Singapore's military was controlled by two limited battalions, Singapore Infantry Regiment (SIR) and a small navy force under the Malayan Naval Forces (MNF) base in Woodlands, and the air force known as the Malayan Auxiliary Air Forces (MAAF). However, these defense forces were under the authority of the Federation of Malaya (Keling, Shuib, and Ajis 2009, 68). More importantly, it was the continued British military presence in Singapore post-World War II that was perceived as essential to Singapore's security and socioeconomic stability – the British troops contributed 25 percent of Singapore's GDP and hired 25,000 locals, which provided a "sense of security, without which we would not get investments and be able to export our goods and services" (Lee 2000, 47). The British used Singapore as a key logistical hub in the Far East, and which later became essential in supporting operations against Indonesian incursions into the Malayan Federation during the 1963–1968 period of *Konfrontasi* (Confrontation) aimed at destabilizing Malaysia. At the time of its independence, Singapore had no expectations that British troops, numbered around 35,000, would be withdrawn or even reduced significantly in the short and medium term – well into the mid-1970s. From 1966 onward, however, Lee Kuan Yew noted a constant uncertainty over the planned duration of British military presence (2000, 49), which began to wind down as a result of the end of *Konfrontasi* and UK's deteriorating economic position from 1968 to 1971 (Huxley 2000, 11). By the mid-1968, the Singapore government concluded with Britain the modalities for the transfer of British bases in Singapore concomitant with the schedule of the withdrawal of the British forces (Republic of Singapore Air Force 2008, 10). On October 31, 1971, the Anglo-Malayan Defense Agreement, a de facto alliance treaty signed in 1957 that stipulated British action in the event of an armed attack on Malaya or any of Britain's remaining territories in the Far East including Singapore, was replaced by a consultative, not binding Five-Power Defense Arrangement (FPDA). In this context, Lee notes, "the old era of underwritten security had ended. From now on we had to be responsible for our own security" (2000, 65). Taken together, the historical circumstances of Singapore's road to independence – the immediate challenges of securing international recognition, defending Singapore, and ensuring economic survival in the absence of any advanced planning – led to the development of a narrative of *perennial sense of vulnerability* (Leifer 2000, 32), which deeply entrenched Singapore's imperative for self-reliance – "as the sine qua non" of Singapore's deterrence and defense against potential aggressors (Huxley 2000, 56; Koh Swee Lean 2014, 119).

The above historical experiences, however, only amplify Singapore's innate geostrategic security predicaments defined by its geographical disposition – location, size, lack of natural resources, and physical limitations of Singapore as a small island city-state of 641 square kilometers – the smallest state in Southeast Asia. On one hand, Singapore's geographic location at the southern end of the Straits of Malacca crosses some of the most important Sea Lines of Communication in the world, linking the Indian Ocean and the Pacific Ocean, which carry a major portion of the world's trade (Koh Swee Lean 2012, 76). The Malacca and Singapore Straits carry more than 30 percent of the world's commerce, half of the world's oil and 80 percent of oil bound for China and Japan (Tay 2007, 14). Strategically, Singapore's position coupled with its deep natural harbor provides a hub for maritime trade in Southeast Asia connecting trade routes between Asia, Europe, America, and the Middle East, which has significantly contributed to Singapore's socioeconomic prosperity. On the other hand, Singapore's strategic location coupled with virtual absence of natural resources amplifies its extreme dependencies on the outside world. In this context, one could argue that Singapore's national security is not threatened as much from a single country, but from the disruption of commerce. According to Huxley, "serious disruption of Singapore's physical links with the outside world would threaten not just its economic wellbeing: its very survival as an independent nation would be at stake" (2000, 32). Singapore's security is, therefore, inextricably linked to the profound strategic interests and involvement of major powers in the region, including risks in the regional security environment as well as shifts in the global economy. S. Rajaratnam noted, "We are valuable in war as a strategic asset. We are valuable in peace as a great commercial center. So whether it is peace or war, we cannot escape the consequences of whatever happens in the Pacific" (Tan 2001, 1). Singapore's strategic location and the attention of Great Powers thus serve as a source of both weakness and strength for Singapore – "it is a negative as it does not welcome the close attention of the Great Powers, but is a strength because it affords Singapore the opportunity of playing off the Great Powers against each other, to its own advantage" (Tan 2001, 1). Indeed, while Singapore has established government-linked companies and foreign investment arms to diversify Singapore's economic dependencies, the sustained character and magnitude of Singapore's inseparable linkages with the world continues to shape both risks and opportunities that determine Singapore's economic survival. Singapore's economic, foreign, and defense policies must, therefore, "balance between regional and international environments to survive both as a world city and a city-state" (Low 2002, 16).

In the military operational perspective, Singapore lacks strategic depth as well as natural resources for sustaining an effective defense against surprise external aggression (Koh Swee Lean 2012, 76), and as in the case of Israel, it can neither trade space for time nor lose a single war. As Goh Chok Tong noted, "the loss of one city would mean the loss of the whole nation" (Chin 1987, 196). In a time of war, the SAF cannot retreat to safe areas and consolidate before counterattacking an enemy, an experience learned in February 1942 when the Imperial Japanese Army overwhelmed the island state's British-led defenses. The lack of depth also decreases early warning time. Singapore's close proximity to potential regional flashpoints and territorial disputes in the South China sea as well as the risks of maritime trade interdiction or transnational crimes at sea, including piracy, play an important role in Singapore's security conceptions. Additional factors such as a high population density, proximity of military and vital installations to residential areas, and lack of safe hinterland for civilian population, all contribute to Singapore's narrative of critical vulnerabilities (Boey 1996) – particularly to air and artillery bombardment that would likely cause devastating casualties. Similarly, in peace time, the SAF faces limited land and airspace, which constraints its infrastructure and training needs relative to the safety and efficiency of both civil and military areas. With Singapore's economic rise coupled with competing use of land requirements for residential, industrial, and development projects, the space available for the SAF has shrunk considerably over the past two decades. Consequently, various units of the SAF, predominantly the army and air force, have to maintain training detachments overseas in friendly countries, including Australia, Brunei, France, Germany, New Zealand, Thailand, Taiwan, and the United States. While SAF overseas training missions and presence may provide an array of benefits – from acquiring a diverse training experiences that may strengthen operational readiness, accelerate technological assimilation process, and benchmark own abilities against more capable militaries, to building sustained defense diplomacy that supports Singapore's long-term strategic interest – the SAF has to continuously seek balance in arranging its limited defense resources and assets along with personnel-intensive activities.

Parallel to the historical legacies and geostrategic conditions of vulnerability, the narrative of Singapore's traditional security paradigm acknowledges asymmetries in demographic factors. Facing two regional powers (and potential threat vectors) of Malaysia and Indonesia, predominantly Muslim-Malays with a combined population of around 285 million (2014), Singapore's population of 5.3 million consisting of 3.2 million Singaporean citizens with majority ethnic Chinese background (77 percent) and 2 million foreign residents had resulted in perceptions of "Chinese island in a Malay sea" (Lee 1998, 15; Da Cunha 2002, 134; Loo Fook Weng 2005, 395). Singapore's own internal ethnic and religious complexities have historically reinforced the city-state's perceptions of vulnerability, particularly through the linkages between political developments in neighboring countries and its spillovers into communal relations in Singapore (Huxley 2000, 33). In this context, Singapore's historical narrative often reiterates the traumatic experiences of the 1950 "Maria Hertogh" riots and the 1964 riots (Ng 2005, 14). Externally, potential ethnic or cultural conflicts coupled with political internal instabilities in neighboring countries could develop into armed conflicts that would threaten Singapore's economic assets. Moreover, Singapore's prevailing dependence on water and gas imported from Malaysia has provided another important vector of security concerns, particularly in the period of strained relations between the two countries in the aftermath of Singapore's independence. At the same time, piracy and illegal immigration in near waters, political crises, and possibilities of renewed internal conflicts in Indonesia have reverberated threats to regional (ASEAN) security and stability. It is important to note, however, that Singapore's political leaders have traditionally refrained from pointing at any specific threats, providing only general statements – that is, Singapore "faces an external environment which is characterized by complexity, turbulence, and unpredictability" (Ho 2006). In rare occasions, one can find more explicit assessments. For example, in 1995 then Singapore's Prime Minister Goh Chok Tong presented a rare and unusually candid overview of potential challenges:

> Several things could go wrong for us. If there is a calamity in Indonesia and there is widespread hunger and strife, thousands of refugees could flood into Singapore. Remember, in 30 years' time, Indonesia would have a population of 300 million. Or our neighbors may do well. With economic strength, they may spend more on their military. If they are well-disposed towards Singapore, we can benefit from their prosperity. If they are hostile, that is troublesome. We have two water agreements with Malaysia: one for 86 million gallons/day (mgd) expiring in 2011, and the other for 250 mgd expiring in 2061. Will Malaysia renew the water agreements in 2061? If so, at what price? If not, what will happen?
>
> (Boey 1996)

The final and perhaps most important circle of vulnerability embraces Singapore's own internal demographic and population trends concomitant with ethnically diversified and relatively small population size, continuing low-level birthrates, the resulting population ageing and smaller talent pool, which have led to fluctuations in the number of males available annually for conscription as Full-time National Servicemen (NSFs). In 2014, Singapore's citizen population of 3.87 million (out of total 5.47 million) grew at its slowest pace in a decade, with a total fertility rate as low as 1.19 (*The Straits Times* 2014). During the 1990s, the SAF enlisted about 15,000 NS men annually. In 2011, the annual intake peaked to 21,000 as cohorts of baby-boomers came of enlistment age, including those born in the 1988 Dragon year. In 2012, Singapore's Ministry of Defense's (MINDEF) long-term projections indicated a gradual reverse to levels of the 1990s, about 15,000 each year, through which Singapore may mobilize about 300,000 soldiers in four divisions consisting of Regulars, NSFs, and Operationally Ready NS men (Ng 2012). Historically, Singapore's defense planners have been continuously searching for both efficient and more productive use of its manpower resources, not only in terms of military expenditures and use of technology but also, more importantly, in terms of organizational learning and leadership experience necessary to maintain Singapore's future governance. Indeed, the SAF has essentially evolved into a core incubator for future public servants and industry leaders through its "SAF overseas scholars" and "Dual-Career Scheme" frameworks, as well as a societal integrator that cultivates a culture of shared responsibility among NSFs, including its core values of professionalism, neutrality, and subordination to civilian controls (Huxley 1993; Raska 2007). Notwithstanding these efforts, the long-term impact of Singapore's demographic changes on security may amplify the problems concomitant with the shortage of active manpower while generating new contending policy debates on issues such as the viability of professional vs. conscript SAF or the integration of women into the SAF.

Singapore's defense strategy: deterrence, diplomacy, readiness, and resilience

Singapore's geostrategic context, historical legacies, and a mix of internal and external security challenges has subsequently shaped the trajectory of its strategic thought, foreign policy and defense planning, and overall force modernization paths. Indeed, the enduring narrative of existential *angst* or extreme vulnerability – "its valuable geostrategic and geoeconomic position in times of war and peace, its small physical size and lack of strategic depth, its problematic relations with its larger neighbors exacerbated by differences in ethnicity combined" (Loo Fook Weng 2004, 363) – has underscored the character of Singapore's evolving defense and foreign policies and continues to resonate to this day, particularly in Singapore's indicative desire for indigenous military capacity and freedom of action. As Bilveer Singh noted, "if lessons can be learnt from history, Singapore made a conscious effort to ensure that history will not repeat itself; that Singapore will not rely entirely on others for its own defense and will not be defenseless in the face of security threats" (2003, 18). In this context, however, the principal challenge for Singapore's defense planners has for nearly five decades reflected the same fundamental problem: how to translate Singapore's limited resources of a small island nation into an effective defense capability amid continuously evolving security challenges?

In the first three years of Singapore's independence, Singapore's external defense was de facto provided by the British military presence. Singapore's indigenous two infantry regiments, commanded by British officers, had considerable limitations; in essence, the SAF had to start building defense capacity from scratch in virtually every aspect of defense planning, organizational force structure, doctrine, training, and weapons procurement. In October 1965, Singapore established the Ministry of the Interior and Defense (MID), headed by Dr Goh Keng Swee as its first minister. This was followed by the opening of the Singapore Armed Forces Training Institute (SAFTI) in February 1966, set up to train officers and non-commissioned officers. With Singapore's small population and competing needs for manpower and economic development, Singapore opted for building a conscript-based citizen armed force against the initial advice of Israeli consultants. On February 27, 1967, the National Service (Amendment) Bill was first introduced in the parliament. It mandated that Singaporean males whose 18th birthday fell on or after January 1, 1967 would be eligible for call-up. Initially, only 10 percent of those eligible for National Service would be selected to serve as NSF as the SAF did not have adequate facilities and only limited number of trainers. This first batch of SAF conscripts would spend two to three years with the army and thereafter 10 years in training the reserves until reaching 40 years of age. The remaining 90 percent would serve part-time, in the so-called People's Defense Force. Singapore's parliament passed the Bill on March 14, 1967, and three days later, the MID in conjunction with the Central Manpower Base began sending registration reminders to over 9,000 young males born between January 1 and June 30, 1949 (Ministry of Defense Singapore 2007, 32). Notwithstanding initial difficulties, particularly in convincing Singaporean-ethnic Chinese to serve in the military, the introduction of National Service led to rapid expansion of the SAF, from 2,000 in 1967, 14,000 in 1970, and 25,000 in 1975 (International Institute for Strategic Studies 1967–1975).

With the exception of national service requirements to its female citizens, Singapore's initial model of compulsory National Service, career army, and reserve service was essentially adapted from the Israel Defense Forces (IDF). From November 1965, only three months into Singapore's independence, Singapore started inviting the first batch of undercover Israeli military personnel, who provided methodological advice, lessons learned, and support guiding the build-up of SAF's military capabilities for the next two decades (Ng 2005, 7). Notwithstanding Singapore's diplomatic requests for military assistance from Egypt and India, which were quietly rejected, Singapore has turned to Israel. At that time, both Prime Minister Lee Kuan Yew and Defense Minister Goh Keng Swee believed that Israel, as a small country surrounded by Muslim countries, with its strong IDF, could help Singapore to build "a small, dynamic army" (Barzilai 2004). Israel's defense establishment recognized the long-term strategic significance of military cooperation (albeit covert) with Singapore and proceeded with the planning. Yitzhak Rabin, then the IDF Chief of Staff, appointed Maj. Gen. Rehavam Ze'evi as Deputy Chief of Staff and Ezer Weizman as Head of the Operations Branch to organize and coordinate the preparatory requirements. Ze'evi then selected Col. Yaakov Elazari to form a team of IDF officers who would be responsible for writing specific chapters for the first SAF military manuals. The first so-called "Brown Book" included chapters dealing with the establishment of an infantry and combat doctrine. This was followed by the "Blue Book" that provided a template for the creation of the Singapore's defense ministry and intelligence units. On December 24, 1965, the first Israeli teams covertly arrived to Singapore and began supervising the establishment of the first SAF military base, which was set up on plans designed by the Israeli Engineering Cops. The Israeli team also began training the SAF through a number of courses at SAFTI, including courses designed for new SAF recruits, platoon commanders, and select officers. In the aftermath of the Israeli victory in the 1967 "Six-Day War," the Singapore–Israeli military ties deepened further, shaping the contours of Singapore's military capabilities, force structure, weapons acquisition, and operational conduct.

In January 1968, for example, Singapore decided to establish an armored corps, which would be organized, trained, and manned based on relevant aspects of Israeli land warfare doctrine. In a highly classified manner, Singapore ordered 72 AMX-13 light tanks from the IDF surplus, which began to arrive in 1969 – and first paraded on Singapore's National Day on August 9, 1969, to the surprise of many foreign diplomats and military observers. At the same time, the IDF also helped to set up the SAF flight school, technical school, and the procurement of a squadron of *Alouette* 3 helicopters along with 40-mm anti-aircraft guns (Barzilai 2004).

In the context of building Singapore's defense management capacity, organizational and operational capabilities, however, it should be noted that Singapore has not directly emulated the Israeli model but only adopted its relevant skills that were adapted to Singapore's unique circumstances, a fact that Israeli advisers also recognized from the onset of their mission. From the Israeli perspective, the IDF was in Singapore to provide the best military advice for the SAF to become self-reliant while recommending the most relevant weapons procurement regardless of its source. On the Singapore side, however, the question of magnitude and impact of Israeli assistance is contested. One school of thought argues that Singapore's strategic objectives and force posture have differed from the Israeli case in the first two decades, as the SAF strategic orientation has been much more defensive, compared to the offensive preemptive armored doctrine of the IDF. Moreover, differences in strategic culture, defense resource availability, and relations and support of super powers, particularly the United States, has dictated a different trajectory for the SAF. As Ng Pak Shun argues, "Singapore neither had the Israeli level of defense spending and the Israeli technical expertise, nor strong lobbies in the United States to emulate the Israeli military doctrines successfully" (2005, 8). Therefore, in the subsequent development, the SAF and its three integrated services – the Singapore Army, the Republic of Singapore Air Force (RSAF), and the Republic of Singapore Navy (RSN) – engaged cooperation with military advisors from other countries, including the United Kingdom, Australia, New Zealand, and others.

At the same time, however, one could argue that the most important contribution of Israeli military mission to Singapore has led to the formation of the first defining strategic pillar in Singapore's defense strategy: *Deterrence*. In the words of Andrew Tan, "the central aim in adopting the Israeli system was to enhance deterrence by proving that Singapore possessed a credible military capability based on a proven model, and would be willing to use this power against any attempted aggression" (1998, 5). Tan essentially presents a counterargument to Ng's analysis, noting that the speed and scale of Singapore's initial military build-up from the late 1960s to early 1970s, which was assisted by the Israeli advisers and enabled by the presence of excellent military facilities set up by the British and significant defense spending taking up some 11 percent of GNP and 40 percent of government expenditures in the late 1960s, far exceeded the goal of a "defense state" (Tan 1998, 6). Indeed, the IDF advisors at that time considered a light armor-based land force component feasible to SAF, based on their own study of the likely area of operations and the subsequent sale of AMX-13s, suggesting at least a limited strategic offensive posture. Bernard Loo argues that Singapore's top priority for "self-reliant" military capacity, and the need to establish a credible deterrence in the SAF's formative years, was driven primarily by Singapore's threat perceptions – the immediate susceptibility to strategic pressures from its larger neighbors, particularly from Malaysia and Indonesia in the early years of Singapore's independence. Malaysian public threats to disconnect water supplies to Singapore, for example, set conditions for the development of SAF land and air power, reflecting the Israeli experience and pattern of defense development (Loo Fook Weng 2004, 367).

From the early 1970s to mid-1980s, Singapore's concept of deterrence was predicated on the development of the so-called "Poisoned Shrimp" strategy, based on an analogy of "easy to swallow, but impossible to digest" (Singh 2003, 26). This "declaratory" concept envisioned the SAF defending the island at its water's edge first, to be followed by an intense close combat in urban areas, with the key strategic aim to elevate the human and material costs of aggression against Singapore to unacceptable levels (Deck 1999, 249). In other words, "while Singapore could not resist a determined invader, the cost of any victory to an invader would be so high as to be an effective deterrent" (Tan 1998, 6). At that time, the strategy was predicated on the limited offensive capabilities of the SAF – its lack of manpower, firepower, and mobility. While the exact operational plans have never been publically specified, it was presumed that the "Poisoned Shrimp" would be augmented with the support of Singapore-based British, Australian, and New Zealand forces (presumably under the FPDA framework), which would assist the SAF in resisting any direct military intervention – whether from Malaysia, Indonesia, or other sources (Huxley 2000, 56). Bernard Loo notes that the "Poisoned Shrimp" strategy was deficient, because "it was essentially defeatist; it envisaged the eventual defeat of Singapore and its disappearance. Its deterrent value was in the promise of great pain that the potential aggressor would suffer in the process of defeating the SAF, but that defeat was virtually guaranteed" (Loo Fook Weng 2004, 367). Indeed, as Brig. Gen. Lee Hsien Loong, then the Chief of Staff of the SAF, acknowledged, "the Poisonous Shrimp strategy was deficient in that it offered Singapore merely a choice of 'suicide or surrender' because of its implication that the SAF would fight an ultimately unwinnable war on its own territory" (Huxley 2000, 57).

In practice, however, with the SAF's accelerating military modernization, equipment, organization, and training throughout the 1970s and 1980s, Singapore's deterrent/ defense strategy increasingly placed an *implicit* emphasis on strategic preemption of potential adversaries – that is, Malaysia. The notion of a preemptive offensive doctrine was initially advocated by the Israeli military advisors during the late 1960s. Given Singapore's geostrategic vulnerabilities and the extreme difficulties of defending Singapore onshore, amplified by regional strategic uncertainties, political instabilities, and robust regional military build-up in the backdrop of the Vietnam War, conventional preemption in broader framework of deterrence "was the only strategy which made sense" (Huxley 2000, 56). At the operational level, the objective of this strategy would aim at achieving two mutually reinforcing objectives:

(1) creating strategic depth to protect the island from direct enemy fire, and (2) moving the forward edge of the battle area in order to secure Singapore's water supplies sourced from southern parts of Malaysia. Bernard Loo argues that "these two primary considerations have driven the evolution of Singapore's military doctrines from essentially defensive to essentially offensive" (Loo Fook Weng 2004, 368). The principal scenarios for a direct military confrontation with Malaysia envisioned an escalating political instability in Kuala Lumpur, widespread communal violence, resulting in waves of ethnic Chinese refugees, and the potential for interference whether by government or nongovernmental forces interdicting Singapore's vital water supplies. According to Huxley, these contingencies have been repeatedly exercised in SAF staff college exercises since the late 1960s (2000, 59). In response, the SAF would rapidly advance into peninsular Malaysia, first by establishing RSAF air power superiority over the Malaysian air force, parallel with infiltrations by SAF commandos to secure the bridges and links to the Malaysian side of the Causeway, followed by maneuver warfare by SAF combined arms forces – that is, armored units equipped with light tanks (and possibly also heavy Centurion MBTs), supported by infantry guards battalions and their logistical elements, and, last but not least, the RSN securing the Singapore and Malacca Straits from Malaysian or other external interference. While operational details on the pace, range, extent, and duration of such SAF maneuver into Malaysia is unclear, informal discussions with informed observers imply a 80-km perimeter zone inside Johor to establish a so-called "Mersing Line" that would ensure Singapore's control of two key water-pumping stations at Skudai and Kota Tinggi, while providing substantial strategic depth to protect Singapore island from the Malaysian artillery (Huxley 2000, 60).

Officially, the shift in declaratory stance in Singapore's defense strategy from the image of "Poisonous Shrimp" to "Porcupine" occurred on January 9, 1982, in a seminal public speech by Brig. Gen. Lee Hsien Loong at the National University of Singapore: "If someone threatens to step on us, and our only alternatives are suicide or surrender, then there will be a very strong argument for surrender. So we need a policy which says: 'If you come I'll whack you, and I'll survive.' This is a workable strategy. I may not completely destroy you, but you will have to pay a high price for trying to subdue me, and you may still not succeed" (Lee 1984a, 164). According to Ng Pak Shun, the change from a de facto *nonexistent* "Poisonous Shrimp" policy to a more offensive and survivable posture reflected a timely maturity of Singapore's military capabilities and its indigenous military planning capacity, as well as the need to project credible defense posture by the SAF, both internally and externally, signaling that "Singapore would, and could, defend itself successfully and survive any potential threat" (2005, 33). In this context, one could argue the predominantly *lexical turn* in how Singapore's leaders articulated SAF's defense posture, represented a gradual shift from the so-called First- to Second-Generation SAF, characterized by increasing, albeit limited, power projection capacity at some distance from its shores (Loo Fook Weng, 70). While, during the 1970s, the build-up of the 1G SAF essentially concentrated on the development of individual services – the quantitative assimilation of select platforms and basic capabilities into the army, air force, and navy – from the mid-1980s, the SAF began experimentation with planning and executing operations at the system level. The shift in emphasis from quantitative platform-centric to qualitative system-level competencies coupled with the drive to strengthen indigenous technological capabilities aimed at raising operational readiness of the SAF (Republic of Singapore Air Force 2008, 52).

In the context of developing defense strategy, however, a significant milestone occurred in 1984 with the introduction of the TD concept. Applicable for both peacetime and wartime, TD as a comprehensive security doctrine has emerged from Singapore's benchmarking of similar strategies in Sweden and Switzerland in the early 1980s – countries with relatively small populations and some form of national military service. TD framed Singapore's security in terms of five mutually reinforcing dimensions – military, civil, economic, social, and psychological security. In doing so, TD also provided a nexus between development, diplomacy, deterrence, and defense – as key strategies for Singapore's survival. For example, in his 1984 speech on "Security Options for Small States," then Brig. Gen. Lee Hsien Loong described this dynamics (Lee 1984b) as follows:

> *Development* is the strengthening of the nation internally by building up a stable and cohesive society, establishing social bonds, encouraging economic growth, and strengthening political institutions ... The aim, as far as security is concerned, is to avoid presenting weaknesses which may be exploited by others, and in extreme cases to avoid self-destruction. Over a period of decades, economic growth ... will translate into potential strength and resilience in every field of endeavour, it will be especially so in defensive power;

> *Diplomacy* refers to the totality of a state's relations with other states in the international system. Self-interest is the governing principle of most nations' behaviour. However, this does not mean that no reliance whatsoever can be placed on other nations, or that nations never come to one another's help ... The only question is therefore whether it is in the enlightened self-interest of other states to help a nation survive. A wise nation will make sure that its survival and well-being are in the interest of other states, so that it will not have to confront its threats and dangers alone.

> *Deterrence* defines security of small nations, as much as that of superpowers. Aggressors are seldom prepared to pay an infinite price in order to subdue a troublesome small nation ... Deterrence cannot be based solely on the threat of murder by suicide. The hope that the threat, when tested, will be found to be empty may tempt adventurous fortune hunters. A credible deterrence must depend on a viable defense, in other words on a strong armed force. The choice confronting the nation, no matter how dire the threat must never be suicide or surrender.

> *Defense* means a small state must therefore fight to win. It cannot win in the sense of overthrowing its enemies and wiping them of the map, but it can defeat the aggressor's force in battle thereby defending the integrity of its

boundaries and compelling a favourable peace … To guard against a conventional military threat, conventional forces are needed forces which can engage the aggressor in a pitched battle and destroy him in mortal combat. To build up such an effective force, small states must mobilize larger proportion of their potential strength than their opponents, which means having a conscript army.

(Lee 1984b)

According to the 1995 MINDEF defense white paper, "Singapore's defense philosophy is not built on the premise of an existing external threat. Instead, Singapore invests in defense to enhance its deterrence capability. As a sovereign state, it must be self-reliant in its defense to prevent threats from arising in the first place. Total Defense is thus the cornerstone of Singapore's deterrence strategy. It unites all sectors of society – government, business, and the people – in the defense of the country" (Ministry of Defense Singapore 1995, 1). TD envisions ongoing mobilization of population and resources to strengthen the readiness, resolve, and resilience of every sector of society as well as government departments – each playing a role in ensuring Singapore's security against all forms of security challenges. Specifically, Psychological Defense refers to an array of educational campaigns (i.e., National Education Programs carried out in schools) aimed at strengthening the collective will and commitment among Singapore's citizens to defend the country; Social Defense, promoted primarily by the Ministry of Community Development, aims at preventing any exploitation of ethnic unrest by stressing the mutual coexistence, cohesion, and harmony based on multicultural consensus and community building regardless of race, language, and religion; Civil Defense, coordinated by ethnically diverse corps of professionals and national servicemen in the Singapore Civil Defense Force, protects and maintains Singapore's civil resources and infrastructure during a national crisis. The objective of Civil Defense is to protect civilian lives, lower the casualty rates, minimize damage to property, and pave the way for a quick return to normalcy. This includes the distribution of essential items and resources such as food, water, fuel, medicine, blood, while also implementing a comprehensive preventive measure to protect the population; Economic Defense has both strategic and operational significance: on one hand, it refers to the contingency planning for the conversion of civilian human resources, technological skills, and capital investments for the military during wartime while ensuring the continuity and function of the economy when these resources are mobilized for war. At the same time, it also recognizes that military power depends on economic strength, and vice versa. In the absence of a strong economy, the costs of creating and maintaining an effective military capability would be too high; Military Defense consists of strong deterrent forces centered around the credible military capabilities of the SAF as well as indigenous defense industrial base that meets SAF's military-technological requirements (Ministry of Defense Singapore 1995, 13–17; Tan 2001, 13; Matthews and Zhang Yan 2007, 380).

In the early 1990s, TD has been briefly embedded into a broader concept of "Total Security," which emphasized the interdependencies of issues and actions in three strategic dimensions – TD, Diplomacy, and Internal Stability. While TD focused on deterrence and crisis management, Singapore's defense policy recognized the need to put an equal emphasis on diplomatic means to ensure cooperative relations with friendly countries, especially its neighbors. "Internal Stability" was conceptualized as a broader umbrella for ensuring social cohesion and harmony. In this context, Singapore's defense planners focused on countering potential internal threats such as internal subversion and racial/ethnic/religious insurrection while simultaneously applying deterrence and diplomacy to prevent interstate conflicts and regional tensions. The 1995 defense white paper described the analogy of a three-dimensional "S-Cube Concept": "Survival, Security, and Success are the building blocks of Singapore's future. The three elements are intertwined inextricably. Survival is the imperative of every nation. But there cannot be survival without security. At the same time, without success there will not be motivation to persevere" (Ministry of Defense Singapore 1995, 1). While the terms underlying these concepts have changed in the subsequent years, the principle of integrating, balancing, and leveraging multiple, in essence, select "soft power" and "hard power" strategies has been a key feature in Singapore's comprehensive approach to its national security.

In the 2000s, Singapore's comprehensive defense strategy continued to evolve around the fundamental tenets of diplomacy (through good international citizenship) and deterrence (through operational readiness and experience) and – should these fail – securing a swift and decisive victory over a potential aggressor. In a 2009 lecture on the "Fundamentals of Singapore's Foreign Policy: Then & Now" at the MFA Diplomatic Academy, Lee Kuan Yew noted, "a small country must seek a maximum number of friends while maintaining the freedom to be itself as a sovereign and independent nation. Both parts of the equation – a maximum number of friends and freedom to be ourselves – are equally important and interrelated" (2009, 1). To achieve this, "we must make ourselves relevant so that other countries have an interest in our continued survival and prosperity as a sovereign and independent nation; we have to be different from others in our neighbourhood and have a competitive edge. Because we have been able to do so, Singapore has risen over our geographical and resource constraints, and has been accepted as a serious player in regional and international fora" (Lee 2009, 1). In the context of its defense diplomacy, the SAF has increasingly focused its activities toward "developing and maintaining good relations with other countries through extensive and expanding links with armed forces in the region and beyond, promoting greater understanding and trust among regional and extra-regional armed forces by leveraging on these links to help strengthen bilateral and multilateral defense cooperation and dialogue, and participating in confidence-building efforts" (Ministry of Defense Singapore 2000, 13). These include close and friendly ties with many of the armed forces of ASEAN countries; bilateral defense relations with countries in the wider Asia Pacific region, including the United States, China, Japan, South Korea, New Zealand, and India; and friendly ties with armed forces in Europe, Africa, and the Middle East. As a part of its 3G "Full-Spectrum Force," the SAF expanded its training toward a nonconventional spectrum of Operations Other

Than War contingencies ranging from Peace Support Operations to Humanitarian and Disaster Relief (HADR). Throughout the 2000s, the SAF also increased the scope of its participation in overseas Stability, Transition, and Reconstruction and HADR operations (Ong 2011, 1), which included SAF's long participation in operations in Afghanistan, Timor Leste, and Gulf of Aden.

Notwithstanding the widening range of its defense diplomacy and other cooperative measures, the SAF's constant improvement and sustained development of its concepts of operations, organizational C4ISR structures, and military-technological capabilities over the past three decades have arguably focused on training and combat-readiness (Tan 1998). Indeed, since the early 1990s, Singapore's defense policy makers and officer corps have debated the impact of new technologies in the context of progressive complexity of East Asia's security challenges, characterized by a convergence of high and low-intensity conflicts. With the introduction of the "3G SAF concept" and "Integrated Knowledge-based Command and Control" (IKC2) doctrine in the mid-2000s, the SAF has invested substantial resources into developing operational and organizational concepts and integrating niche advanced weapons platforms and systems, including satellite capabilities, airpower and precision weapons, and networked targeting. Conceptually, Singapore's defense strategy has shifted again in the 2000s from the analogy of a "porcupine" – aimed at increasing capacity to project at least limited military power (the porcupine's quills) at some distance from its shores – to a "dolphin" that uses intelligence, speed, and maneuverability in a spectrum of diverse missions: from defense diplomacy and operations other than war to kinetic precision strike capabilities conducted further afield from the immediate environment of Singapore (Loo Fook Weng 2015, 70).

3G SAF: conceptual challenge and adaptation

Singapore's evolving security and defense policy, anchored around a remarkable continuity of Singapore's traditional security paradigm, has been reflected in the direction, pace, and character of Singapore's military modernization as well as its aspirations for an indigenous military innovation: the 1G SAF (1960s–1970s) focusing on capability development of individual services; the 2G SAF reflected a period of consolidation and adaptation from service-oriented strategic thinking toward conventionally oriented combined arms warfare (1980s–1990s); and the 3G SAF (2000s onwards) synonymous with a transition toward a joint strategy leveraging both defense diplomacy to enhance peace and security in the region, as well as RMA-oriented strike capabilities in the SAF. In this context, while the 1G SAF inherited and adapted technologies, operational and organizational concepts from other military traditions (i.e., IDF), the 2G SAF focused on adapting these concepts and technologies to Singapore's defense requirements. With the ongoing 3G force, however, the SAF has aimed to develop its own indigenous technologies, concepts of operations, and organizational structures to sustain its regional "strategic edge" in an increasingly uncertain regional security environment (Chan *et al.* 2005, 1). Indeed, for over a decade, the SAF's 3G force transformation has been conceptualized not only in terms of new technologies but also primarily as a fundamental conceptual, organizational, and doctrinal change (Lim 2003). In doing so, SAF's primary mission statement has reflected both continuity and change in its overall strategic orientation: to protect the interests, sovereignty, and territorial integrity of the Republic of Singapore from external threats; deter potential aggressors from attacking Singapore; and, if deterrence and diplomacy fails, to secure a swift and decisive victory (Ministry of Defense Singapore 2000). In 2014, MINDEF modified its mission statement for the SAF by adding a new role "to enhance Singapore's peace and security," which implied a nuanced strategic shift in the use of "soft power" by the SAF (defense diplomacy) on top of its core military capabilities.

Seen from this perspective, it is imperative to outline the essential contours of intellectual history and concepts that have shaped the character and process of Singapore's 3G force transformation. Officially, the concept of a "3G SAF" was first publically unveiled in a "Statement at the 2004 Budget Debate in the Singapore Parliament" by then Minister for Defense Teo Chee Hean on March 15, 2004. In the detailed speech, Teo noted the progressive complexity and challenges facing Singapore's security environment, particularly the increasing relevance of nonconventional security threats, which have widened the spectrum of missions of the SAF (i.e., peacekeeping, counterterrorism, and counterproliferation). Coupled with the ongoing RMA-related technological advances introduced at the conventional level of warfare, particularly in the Iraq War of 2003, Teo called for the need for the SAF to develop future-oriented capabilities beyond maintaining the current operational readiness. Specifically, Teo (2004) noted:

> The advantage in the battlefield of the future lies with the force that can harness technology to make maximum use of these transformational capabilities. Platforms, manned and unmanned, weapons and sensors that are fully networked into such a fighting system will have their combat power magnified many times. Tanks or ships or aircraft operating on their own, which are not so networked and which are capable of operating only in the conventional way, may find that they are merely targets.
>
> The transformation of the SAF to exploit these rapidly emerging possibilities is a strategic imperative. But, it is not something that can be achieved overnight. To introduce new concepts of operations and new technologies requires experimentation and research and development. It will ultimately lead to changes in organization, less demand for conventional platforms, more demand for less visible technologies like information system, precision weapons, electronic warfare systems, unmanned platform technologies, and a new type of soldier who is trained to exploit these capabilities. This is a major and strategic effort we have already embarked on, but it will take time.
>
> … While MINDEF will continue to modernize the SAF, such as through the acquisition of the Next Generational Fighter, the effort in transformation will eventually see a *third generation SAF* emerging over the longer term. This will be an SAF that has enhanced capabilities through exploitation of new concepts and technologies, an ability to fight across a wider spectrum of operations, flexible, and much more efficient in the use of resources.

Within the SAF, however, the conceptual development of network-centric warfare and concomitant RMA debates has been studied and explored since the mid-1990s. At that time, Singapore's defense establishment reflected on the U.S. experiences in the Gulf War and later the Kosovo Conflict, noting the emerging RMA debate in the United States and subsequently debating its applicability and impact on the SAF. This is evident in the number of articles on the RMA published during the 1990s by the *Pointer – Journal of Singapore Armed Forces*. For example, article titles such as "Application of Advanced Military Technology in Desert Storm" (Leong Sek Kay 1995); "Revolution in Military Affairs and Command, Control, Communications and Computers (C4): Are We Ready?" (Lim 1998); "The Impact of Technology on the Military: An SAF Perspective" (Chen 1999); "Joint Vision 2010: The Concept of Future Warfighting for the US Armed Forces and its Relevance to the SAF" (Tay 1999) explored the strategic implications of advanced information technologies in combat and, more importantly, its impact on the SAF. The majority of views within the SAF at that time viewed the RMA as a paradigm shift bringing about technological advances that would enable the SAF to overcome its resource and manpower constraints and, in doing, ensure SAF's future "strategic edge." In 1995, for example, Brig. Gen. Han Eng Juan declared that the coming technological revolution would change the way battles would be fought:

> Electronic connectivity among all the echelons of the army in a digitized battlefield will result in the speed and precision never before seen in military maneuver. The entire army's Command, Control, Communication and Computer and Intelligence system will be revolutionized. In the future high-tech digital battlefield, information may be even more critical than ammunition. With superior information technology, the army's situational awareness and agility will be greatly enhanced, resulting in significantly improved lethality, survivability, speed, versatility and sustainability.
>
> (1996, 20)

In this context, however, the SAF recognized the necessity to go beyond emulation of select foreign military doctrines and operational concepts. For example, in the preface for the book commemorating the opening of the new SAFTI Military Institute in 1995, then Chief of Defense Maj. Gen. Bey Soo Khian wrote, "the organization and technology of warfare have altered dramatically and the SAF officer of the 21st century will encounter the complexities of integrating men, technology and tactics in the fine art of war. To meet this challenge, *we can no longer rely on borrowed doctrines and methods*. We must develop home-grown doctrines and strategies suitable for our environment and compatible with locally-modified weapon systems and indigenous organizational structure" (Menon 1995). To do so, the SAF embarked on intellectual development of its RMA-oriented concepts that became embedded into the broader 3G force transformation processes. This can be inferred from the publication of three SAFTI Military Institute's *Pointer* monographs: *Creating the Capacity to Change* (2003); *Realizing Integrated Knowledge-based Command and Control – IKC2* (2003); and *Spirit and System: Leadership Development for a Third Generation SAF* (2005). Each has subsequently provided a frame of reference as well as a set of ideas that facilitated transformational discourse and internal debates within the SAF on its future direction.

The first monograph, *Creating the Capacity to Change* (*C2C*), represented a unique document for its time, designed to stimulate internal debate within the SAF and, in doing so, inject continuous adaptability into its structure and culture (Ho 2003). To begin with, *C2C* noted that "uncertainty and the rapid pace of change suggest that the SAF should not tie its future development to linear projections from the present. Instead, the SAF must expect and assume that there will be nonlinear disruptions to the strategic environment, in operational concepts and in technology. This demands that the SAF develop a capacity to change and innovate quickly in anticipation of, or in response to, emerging trends" (Choy *et al.* 2003, 2). Such nonlinear disruptions would emanate from the convergence of three sets of variables: (1) *technological discontinuities* in genetics, biometrics, computing power, broadband data communications, GPS, precision-guided munitions, robotics, nanotechnology, (2) *emerging effects of asymmetry* and disproportionality in the rise of nonstate actors, and (3) the *amplifying effects of globalization*. Consequently, the monograph argued,

> [These] trends lay the foundation for any number of radical changes, and indicate a very real possibility of fundamental surprise – by events that fall outside our very conceptions of security.
>
> For MINDEF and the SAF, it is crucial that we learn to recognise when and where our existing planning tools are deficient. While our ability to solve complicated technological and operational problems is unquestioned, more serious doubts exist about our capacity to cope with the complexities generated by a changing strategic environment.
>
> (Choy *et al.* 2003, 3)

By openly acknowledging serious deficiencies, not only in the SAF's defense planning processes but also, more importantly, in the SAF's strategic culture, *C2C* called for a greater organizational capacity to change, aimed to mitigate MINDEF's traditional bureaucratic inertia embedded in its organizational resistance to change and, in doing so, streamline Singapore's strategic and operational adaptability to cope with unpredictable events or crises. "The organizational challenge for MINDEF and the SAF is to nurture a hotbed of truly innovative ideas alongside our regular, well-functioning bureaucracy. There must be space for both change and continuity, for both are important to the success of the organization" (Choy *et al.* 2003, 12). In order to generate *C2C*, the authors proposed ideas based on theoretical template of "complex adaptive systems": "The key organizational lesson from complex adaptive systems is that leaders and managers should find ways to allow creativity to emerge naturally within organizations rather than impose preconceived solutions on their followers and employees" (Choy *et al.* 2003, 16). In other words, military innovation in the SAF must occur, first and foremost, in a credible environment that supports learning and innovation – by building on existing and new processes that would enable the streamlining of relevant new ideas to be recognized and followed through into experimentation and implementation. To do so, the SAF should empower its people with the rights skills to take advantage of this creative space whether in conceptualizing strategy, selecting capabilities, and using force. One of the key recommendations, according to the authors, would be to foster diversity in the SAF's strategic culture: "widen the range of our sources of analysis and cultivate even those diametrically opposed to our conventional wisdom. In addition, actively nurture people within our own organic pool of talent who represent diverse perspectives" (Choy *et al.* 2003, 43).

Following the publication of the *C2C* framework, which called for a more fundamental conceptual and doctrinal change, Singapore's RMA-oriented narrative has evolved significantly further with the publication of the second *Pointer* monograph titled *Realizing Integrated Knowledge-based Command and Control: Transforming the SAF*. Published in 2003, the monograph introduced the contours of the principal operational concept underscoring the 3G SAF: IKC2. In particular, IKC2 has taken its cue from the U.S. strategic thought, initiatives, and operational experience with "Network-Centric Warfare" and "Effects-based Operations" (Ho 2009). It envisioned the networking of sensors and forces by extending the information domain and creating conditions for the emergence of network-centric operations. In other words, IKC2 implied a symbiotic relationship between shared information, increased shared situational awareness, and increased combat power. The authors defined IKC2 as

> Integrated warfare, network-enabled, organized around knowledge for effective command and control … It builds on the comparative advantage of the SAF in its techno-savvy people, technologically superior forces and a systems approach. As such, it is a force multiplier that will enable us to do more with less through the overlapping of sensors, shooters and communication nets across the entire battlespace.
>
> IKC2 is conceptualised as network-enabled, knowledge-based warfighting that is predicated on the OODA loop. "Observing", "Orienting", "Deciding" and "Acting" are essential components of any war-fighting cycle from the way information is assimilated, decisions made and actions taken. IKC2 therefore aims to exploit C4IT technologies to ensure that not only does one's OODA operate faster than the adversary's, but also that we are able to effectively disrupt the adversary's OODA cycle.
>
> (Lee *et al.* 2003, 13)

In this context, IKC2 mapped strategic interactions between cognitive, informational, and physical domains in future combat and defined a myriad of new lexical terms: Pervasive Battlespace Awareness (PBA), Superior Battlespace Understanding (SBU), Knowledge-enabled Decision Superiority (KeDS), and Dominant Battle Management (DBM). Specifically, the PBA located in the physical domain would enable the networking of all sensor resources and, in doing so, enhance the shared situational awareness among forces at all levels and across all services – under the principle of "See First, See More." The PBA would link directly into the SBU located in the cognitive domain and provide commanders and decision makers an in-depth understanding of the battlespace through the fusion, correlation, and prioritization of existing knowledge across networks and real-time contextualized intelligence (understand better and faster). The KeDS also in the cognitive domain would directly augment the SBU by streamlining automated decision support systems (decide better and faster) – that is, by analyzing and proposing creative options for mission planning, monitoring the progress of operations, and assessing mission outcomes. Ultimately, DBM in the physical domain would enable cooperative engagement capabilities (acting decisively) by real-time dissemination of commanders' intent to subordinate units, dynamic tasking and retasking of platforms, weapons, and units based on changed target priorities and data and knowledge acquired from the PBA. The matrix integration of these varying capabilities would effectively multiply SAF's possibilities for future "Integrated Warfare" envisioned in concepts such as the Strategic Sensor Web; Theatre-Wide Precision Strike; Cooperative Engagement Capabilities; and Just-in-Time Automated Logistics (Lee *et al.* 2003, 13–18). In order to implement IKC2, the monograph argued that the SAF needs both a "top-down" strategy and governance to build an "enterprise architecture" essential to achieve an RMA-oriented "system-of-systems" as well as "bottom-up" experimentation approach across the services (Lee *et al.* 2003, 18).

The third influential study shaping the evolving strategic thought of the SAF's 3G transformation focused on the imperative to strengthen SAF's leadership development (LD) and organizational learning, building upon existing foundations of professionalism, moral and ethical principles within the SAF. In 2005, SAFTI Military Institute published a monograph titled *Spirit and System: Leadership Development for a Third Generation SAF*. Its authors argued that "3G SAF will require not only experimentation with new technologies and concepts but also a certain kind of leadership in the SAF – one that is highly adaptive, innovative, and able to cope well with uncertainty and change" (Chan *et al.* 2005, 1). In particular, with the changing strategic context and widening operational spectrum of the twenty-first century, the authors called for enhancing SAF's LD in two interrelated domains: systems and spirit. They defined "Systems" as "the doctrines, curricula, methods and tools that provide a common language and approach to drive the thinking on leadership and leadership development processes in SAF Units and Schools," while "Spirit" refers to "the ethos and motivation of SAF Leaders that drive their everyday actions in the SAF, and is a particularly important element of effective LD." In this context, the authors argued that "LD system implemented without the right Spirit on the part of SAF Leaders will not work. Professional credibility and authority are vital aspects of leadership development. SAF Leaders must therefore always 'walk the talk' and anchor their actions and decisions on the values and purpose of the SAF" (Chan *et al.* 2005, 35).

Historically, the emphasis on leadership, integrity, learning, and professionalism has been at the SAF's core since its inception. Indeed, the SAF has not been born in battle, nor has it honed a heroic military tradition. In 1967, the political establishment formulated a "Code of Conduct for the Armed Forces," which has shaped the values and norms for the behavior, position, role, and functions of the 1G and 2G SAF. At that time, Dr Goh Keng Swee, Singapore's first minister of defense, emphasized that in the absence of a professional soldier caste, a code of guidance was necessary to provide a sense of dignity and purpose to the SAF. The Code stipulated a strictly professional role for the SAF as an instrument of the state in defending its security. It called for all SAF members not to interfere with party politics, remain loyal to the government at any time, be courageous, disciplined, moral, decorous, and dedicated at all times. In 1986, the SAF began to conceptualize a common value system, a process that culminated in a decade later in 1996 with the establishment of seven core values applicable for all ranks within the SAF: *Loyalty to Country*; *Leadership*; *Discipline*; *Professionalism*; *Fighting Spirit*; *Ethics*; *and Care for Soldiers* (Chan *et al.* 2005, 56–64). In 2013, the SAF added an eighth core value *Safety*. In short, the SAF has emphasized a high degree of leadership, self-control of behavior though codes of ethics, a meritocratic system of rewards based on intellectual achievement and integrity. However, the imperative to facilitate "spiritual-ethical" aspects in both LD and organizational learning domains became conceptualized as increasingly important enablers for the 3G SAF transformation.

Organizational and operational implementation

Notwithstanding the above intellectual endeavors to "deconstruct" elements required for military innovation relevant for the SAF – that is, strategic culture, concepts of operations, technology, leadership and organizational learning – its actual implementation has been shaped directly in an ongoing process of gradual defense management reforms, organizational and operational adaptation within the MINDEF itself. To begin with, MINDEF conceptualized the implementation of the 3G SAF transformation into a *three-phase building block approach*: In the first phase, the SAF would acquire new equipment, introduce progressively more capable systems, and establish new units to enable the transformation of the SAF into an advanced, networked force. In the second phase, the SAF would set up new operational commands relevant with an expanded spectrum of operations, at home and overseas, and, in doing so, focus on widening its operational flexibility and responsiveness. And in the third phase, the SAF would aim on enhancing its leadership and human capital through the introduction of enhanced career schemes as well as revision of training and curriculum to maintain a steady stream of capable and committed people to meet the requirements of the 3G SAF (Teo 2010a). Arguably, phase 1 has been the most visible: over the last decade, Singapore has invested in new types of advanced weapon platforms across its air, land, and maritime domains, ranging from submarines, stealth frigates, new-generation fighter aircraft, main battle tanks, artillery, and command, control, communications, and intelligence infrastructure. Some of the most important acquisitions included the *F-15SGs* multirole fighters, *Gulfstream* airborne early-warning aircrafts, *Sikorsky* S-70B naval helicopters, *Leopard 2A4* main battle tanks, *Terrex* infantry carrier vehicles, high-mobility artillery rocket system, *Apache* helicopters, and *Formidable-class* multirole stealth frigates. At the same time, the SAF has progressively replaced existing 2G SAF systems: for example, in 2011, the RSAF replaced the *Searcher*-class unmanned aerial vehicles (UAVs) in service since the 1990s with the *Heron 1* UAVs, which provide longer flight endurance and enhanced surveillance capabilities. The RSAF has also replaced its 30-year-old *RAPIER* air defense system with *SPYDER-SR* ground-based air defense system. The RSN began designing a new class of modular *Littoral Mission Vessels* to replace its 2G *Fearless*-class patrol vessels, also in service since the 1990s (Teo 2011). In November 2013, Singapore announced a contract with German shipbuilder ThyssenKrupp to acquire two advanced *Type-218SG* submarines designed for littoral, shallow sea operations (Raska 2014).

In this context, Singapore's military modernization strategy has broadly evolved along two major paths: upgrading of existing equipment while selectively introducing new-generation systems designed to transform the SAF. MINDEF conceptualized this strategy along three track vectors: (1) *Force Readiness*, (2) *Force Evolution*, and (3) *Force Transformation* (Lim 2003). Force readiness has been viewed as the baseline ability of the SAF to maintain a high level of operational readiness – that is, the ability to rapidly transform from a disciplined and well-prepared force during peacetime to an effective fighting force during wartime, capable of conducting diverse missions in an expanded spectrum of operations. Force evolution has been conceptualized as systematic development of existing capabilities based on the current conceptual and doctrinal frameworks focusing on short- and mid-term requirements. Meanwhile, force transformation has aimed at creating capacity for a long-term conceptual, organizational, and capability transition based on continuous experimentation, organizational learning, and operational experience that would leverage "cutting-edge" technologies in multimission capable forces. In this context, one could argue that the SAF has maintained a consistent "spiral" capability development in areas central to its future defense strategy: stand-off precision technology, protection technology, stealth technology, unmanned technology, information/cyber technology, enhanced lift and endurance, coupled with advanced computer modeling and simulations (Ministry of Defense Singapore 2000).

In the process, however, the principal challenge for the SAF has not been in the defense management – develop, buy, or modify sophisticated "hardware" adapted for Singapore's requirements – but, more importantly, in developing the "software" – devising organizational force structures and operational concepts that would enable the ability to effectively network these systems and capabilities without compromising operational readiness using fewer people while outsourcing select logistics and support functions. In other words, *how to transform the SAF into a force that is flexible, capable to deal with uncertainty, and able to do more with less.* Since the early 2000s, MINDEF has prioritized creating relevant structures and processes to explore new ideas and concepts that would strengthen inter-service and inter-agency integration. In 2007, for example, as part of the phase 2 of 3G SAF, the RSAF has undergone a fundamental restructuring of its existing Commands, which have been revamped into five new Commands (Air Defense and Operations Command, Air Combat Command, Participation Command, Air Power Generation Command, and Unmanned Aerial Vehicle Command). In early 2009, the RSN has also restructured its previous Coastal Command into Maritime Security Task Force that brought together national maritime agencies including the Police Coast Guard, the Maritime and Port Authority, and Singapore Customs to achieve greater operational flexibility and responsiveness to potential maritime security threats. In 2009, the SAF also formed the Special Operations Task Force that similarly integrated diverse units from the elite special operations community (Teo 2010b, 17). More recently, in 2013, the SAF established its Cyber Defense Operations Hub, which consolidated existing SAF units tasked with computer network operations (Chow 2013).

In terms of implementing SAF's long-term transformation strategy, it is important to briefly map its origins, direction, and character. One of the principal policy architects in the early phases of the SAF's 3G transformation process was Peter Ho, then the Head of Singapore's Civil Service and Permanent Secretary for Defense. Ho envisioned a 3G SAF transformational change as an ongoing process in which "an organisation must look forward to the future as an ever shifting horizon, identify relevant challenges, and change to meet these challenges. It must be geared for continuous change. Such change must go beyond mere incremental improvement. It must be transformational, in the sense that when implemented, it will transform the capability of the organisation to fulfil its core mission" (2006, 2). In retrospective, Ho recognized that the existing planning and operations of SAF are caught up in short-term day-to-day operational requirements that preclude long-term strategic thought about transformational change – as opposed to incremental change. In his 2003 speech at a launch of the *Creating the Capacity to Change*, Ho shared his views on pursuing military innovation within the SAF. His remarks provide valuable insights on how Singapore's senior defense policy makers at that time viewed the 3G SAF transformation – not as a disruptive change but rather as an *ongoing process of balancing* – between current and future defense requirements, strategic adaptation and operational experimentation, evolutionary modernization and disruptive innovation, and concomitant resource allocation:

1   Pursuing transformation is a continuous process. There is no end point. It is a journey without a fixed destination. An organization must look forward to the future as an ever-shifting horizon, identify relevant challenges, and change to meet these challenges. It must be geared for continuous change. Such change must go beyond mere incremental improvement. It must be transformational, in the sense that when implemented it will transform the capability of the organization to fulfill its core mission.

2   Transformation does not mean changing the entire organization. At any time, 10–15 percent of forces ought to be involved in transformation. The point is that we do not have to, nor should we aim to, change everything at once.

3   For an organization to transform and innovate, it must learn the skill of balancing opposing imperatives. For example, we want to focus on operational

excellence and incremental improvement to satisfy current requirements. Yet, if these imperatives are not balanced with the ability to master disruptive change and expeditionary thinking, we will not have a future.

4   Get the operational concepts right. Technology is then merely the enabler. It goes without saying that technology is critical and strategic. But do not believe that technology will solve all problems in the future or that high-tech solutions are inherently innovative and transformational.

5   To truly transform and innovate, we must be willing to embrace the unknown. By definition, transformation means journeying into the unknown without a map like Magellan, Captain James Cook, and other great maritime explorers.

6   We must be prepared to experiment and to learn from failure. While the traditional R&D model focuses on "what we know we don't know," our experimental processes must also be geared toward addressing the "unknown unknowns." To experiment, we must shift our cultural values, accept and even embrace reasonable levels of risk. And we must learn from failure.

7   Experimentation and thinking about the future are processes that ought to be insulated from the demands of day-to-day operations. Our existing planning and operations staff are so caught up with firefighting and other day-to-day matters that they have little spare bandwidth to really delve into thinking about transformational change – as opposed to incremental change.

To implement 3G SAF, Ho argued that the MINDEF needed an "organizational equivalent for the individual maverick" that would serve as a "catalyst for transformational change within the SAF" given the "growing organizational complexity of MINDEF and the SAF" (2003, 1). He envisioned the establishment of a small unit with a mandate to experiment with new ideas, technologies, and concepts that would "test the boundaries" – the Future Systems Directorate (FSD). In this context, MINDEF formally launched the FSD on November 5, 2003, tasked to promote innovation, experimentation, and radical change within the SAF. Essentially, the FSD would focus on military innovation by "developing advanced and alternate operational game-changing concepts and disruptive technologies that would 'stress-test' the mainstream ideas, and deliver cutting-edge capabilities for the SAF" (Chan 2015). At the same time, the FSD would think about the long-term strategic challenges facing the SAF, develop new operational concepts, experiment with these concepts, and then implement them (Ho 2003). For its inaugural year in 2004, the FSD was given 1 percent of Singapore's overall defense budget (est. S$8.25 billion) to generate ideas outside the mainstream planning system and to challenge the system with different views (Khoo 2004).

Building on the organizational reforms in the mid-2000s, MINDEF also established the Defense Research and Technology Office (DRTO) in 2006, bringing together existing research and technology (R&T) planning and management units under a single agency, and centralizing key responsibilities for R&T master planning, management of R&T portfolios, and ultimately outcomes. In retrospect, the creation of both FSD and DRTO could be seen as a natural progression in Singapore's continuous efforts to leverage science and technology as a force multiplier to overcome the constraints of Singapore's small size and limited resources. Indeed, as early as in 1971, MINDEF under its first defense minister, Dr Goh Keng Swee, established the "Electronic Warfare Study Group" code-named "Electronics Test Center" (ETC). According to Teo Chee Hean, former defense minister, the small group in the 1G SAF ETC served as a foundation for the development of Singapore's advanced "defense technological ecosystem," consisting of 5,000 defense scientists, engineers, acquisition professionals, and logisticians working in MINDEF, the SAF, the Defense Science and Technology Agency (DSTA), DSO National Laboratories, defense industries, research institutes, and academia (Teo 2008). The SAF often refers to the Singapore's defense technological community as its "fourth service" in addition to the army, the navy and the air force (Bin Osman 2014).

In July 2013, MINDEF merged both the FSD and DRTO, forming the Future Systems and Technology Directorate (FSTD) as the "logical next step to structurally entrench the operational and technological synergies (Ops-Tech) of these key roles, supported by a larger critical mass of Ops-Tech talents and experts … and to drive innovation in concept development and technology exploitation to the next higher level and make FSTD the center of gravity of Ops-Tech integration" (Chan 2015). In doing so, the FSTD fused operational and technological expertise in pursuing military innovation relevant for the SAF. Over the years, FSTD's "Future Systems Architects" including Brig. Gen. Jimmy Khoo, Brig. Gen. Tan Yih San, Rear-Admiral Harris Chan along with SAF's chief defense scientists, including Quek Tong Boon, continued to shape SAF's strategic thought on future operational paradigms, technological force multipliers, scalable platforms, unmanned systems, command and control imperatives beyond the 3G SAF. While most of its activities, accomplishments, and contributions remain clouded in secrecy, the FSTD has been credited with the initial conceptualization for the Airspace Management Technology system that regulates dense aerial traffic over Singapore by assessing the vertical, lateral, and time dimensions of airspace and synthesizing the calculations into one system to help air traffic controllers obtain better situational awareness in and around Singapore's airspace. Other publically known examples of FSTD joint projects with the DSTA and DSO include the development of the Combat Management System onboard the Republic of Singapore Navy's stealth frigates, which detects, tracks, identifies, and prioritizes contacts, and assigns weapon systems to engage enemy targets facing the ships in a seamless "sensor-to-shooter" integration (Teo 2008). Arguably, however, one of the most important "network-enabled force" R&D programs, bringing together innovative operational concepts and technologies from the SAF, DSTA, FSTD, and local defense industry, has been the Advanced Combat Man System (ACMS). Launched in 1998, the ACMS program aimed to "leverage cutting edge technologies as force multipliers to increase the SAF soldiers' C4I, lethality and survivability … by symbiotically linking soldiers to soldiers, platforms and sensors, via a suite of wearable computer tactical communications and integrated weapons systems on the soldier" (Kong 2007, 88). In many ways, the spiral development of the many technologies in the ACMS program could be seen as a Singaporean adaptation and experimentation based on the Israeli "Tzayad" Digital Army Program and Integrated Advanced Soldier program, designed to provide individual soldiers with capabilities for increased situational awareness ("Blue Force Tracking," "Red Force Marking," remote-sensor information sharing) as well as integrated weapon capabilities ("Call-for-Fire" – coordinating and designating precision fires against selected targets).

In recent years, the FSTD has focused more on the long-term strategic perspectives and visions of the future warfare based on the study of emerging operational concepts and their applications in diverse conflicts. In 2015, for example, the FSTD supported the MINDEF in mapping emerging operational concepts and strategies concomitant with "hybrid warfare" – characterized by "an orchestrated campaign to fracture the solidarity of the target nation through undermining its defences in civil, economic, social, psychological and military spheres" (Ng 2015). Studying lessons learned from recent conflicts in the Middle East and Eastern Europe, Singapore's defense planners have expressed concerns about its potential diffusion, adoption, and adaptation in the East Asian security context – situations where conflict may be ambiguous – such as in the context of territorial disputes in the East and South China seas in which armed fighting may not yet have occurred, but the war is already raging psychologically, politically, and economically. The battle may be so subtle and incremental, and the propaganda war so abstruse, that a state may not know it is even challenged (Raska and Bitzinger 2015). Arguably, hybrid warfare has propelled the MINDEF to conceptualize the SAF transformation beyond the current 3G models. On one hand, MINDEF began to envision the contours of the so-called "SAF 2030 Force" in which future systems that are currently prototyped or envisioned may become operational. In 2015, for example, Dr. Ng Eng Hen, Singapore's defense minister, noted possibilities such as the use of multiple micro-UAVs for individual soldiers; robotic mules that can carry very heavy payloads and follow soldiers autonomously; multirole robotics; unmanned ground vehicles; unmanned surface vessels; cyber defense units; and other technologies that will be integrated into existing networked systems that will amplify SAF's intelligence capabilities, as well as mobility, agility, and lethality. In other words, the 4G SAF would leverage a multilayered defense network of integrated sensors and shooters across all maritime, air, ground, and cyber domains with increasing emphasis on unmanned systems and platforms (Ng 2015). At the same time, however, the SAF's future "kinetic strike capabilities" would have to be complemented by greater societal cohesion and resilience, seen as fundamental to the social, economic, and psychological pillars of TD (Ho 2015).

Ultimately, while the concept of 4G SAF may emanate patterns of disruptive change, in practice it reflects SAF's continuous evolutionary adaptation trajectory, augmented by relatively gradual military modernization path shaped by the same strategic objective: to ensure Singapore's security, sovereignty, autonomy, and independence through deterrence, defense, readiness, and resilience in tackling Singapore's core strategic vulnerabilities (lack of strategic depth, small population, and unique position), as well as a range of potential future "hybrid" threats. However, herein lies Singapore's principal and enduring defense management challenge – the question of how to strike a balance between preserving tried-and-tested strategies and structures with finding innovative operational concepts, technologies, and organizational structures in preparation for multilevel conflicts or complex cascading crises. In other words, the key dilemma facing Singaporean defense planners is the question how to build a force and doctrine capable of dealing simultaneously with current security threats while anticipating future challenges. Indeed, at the time of this writing, there seems to be only limited debate within Singapore's defense establishment (and virtually no external debate), arguably given the SAF's ability to maintain its present operational readiness and military-technological "strategic edge" vis-à-vis other militaries in the region. However, in the long term, the SAF may have to rethink its traditional conceptions. In particular, given the increasingly contending strategic realities in East Asia, driven primarily by the rise of China and resulting great power competition played out in the territorial disputes in East and South China seas, the accelerating regional military modernization drive and diffusion of power projection capabilities, the confluence of conventional and nontraditional security challenges in the region – all these factors will have implications on the future of the SAF. At the same time, Singapore's internal demographic constraints – persistently low birth rates, aging population, and the resulting broader socioeconomic changes and challenges – will inherently also have an impact on the future of SAF. While Singapore's defense planners are arguably well aware of these challenges, open debates questioning Singapore's future defense trajectories are virtually absent. Should the SAF remain focused in its core mission capabilities or continuously expand its strategic orientation toward operations other than war? Should the SAF experiment with professional (voluntary) organizational force structure needed for more technologically advanced military or remain anchored in traditional conscript model designed to strengthen sociocultural integration of Singapore's diverse ethnic population? These questions may provide a greater strategic imperative for the SAF to pursue a more disruptive type of military innovation. As Bernard Loo (2015) noted, "for the SAF to remain strategically relevant, it has had to grapple with the fundamental questions of what its mission is, the security challenges that it is likely to face (at any given moment), and the context (both external security and domestic political) in which the SAF will find itself."

References

Barzilai, Amnon. 2004. "A Deep, Dark, Secret Love Affair." *Ha'aretz*, July 16.

Bin Osman, Mohamad Maliki. 2014. "Speech by Minister of State for Defence, Dr Mohamad Maliki Bin Osman, at the Young Defence Scientists Programme Congress 2014." Ministry of Defense, March 25, 1–3.

Boey, David. 1996. "Singapore: A Fragile Nation Toughens Up." *Jane's Intelligence Review*, August 7.

Chan, Harris. 2015. "Message From FSTA." Ministry of Defense Singapore, March 19.

Chan, Kim Yin, Sukhmohinder Singh, Regena Ramaya, and Kwee Hoon Lim. 2005. *Spirit and System: Leadership Development for a Third Generation SAF.* Singapore: SAFTI Military Institute.

Chen, Yuanxin. 1999. "The Impact of Technology on the Military: An SAF Perspective." *Pointer – Journal of Singapore Armed Forces* 25 (2).

Chin, Kin Wah. 1987. "Singapore: Threat Perception and Defence Spending in a City State." In *Defence Spending in Southeast Asia*, edited by Kin Wah Chin, 194–224. Singapore: ISEAS Publishing.

Chow, Jermyn. 2013. "SAF Sets Up New 'Cyber Army' to Fight Digital Threats." *The Straits Times*, June 30.

Choy, Dawen, Ju-Hon Kwek, Chung Han Lai, *et al*. 2003. *Creating the Capacity to Change: Defence Entrepreneurship for the 21st Century*. Singapore: SAFTI Military Institute.

Da Cunha, Derek. 2002. *Singapore in the New Millennium: Challenges Facing the City-State*. Singapore: ISEAS Publishing.

Deck, Richard. 1999. "Singapore: Comrehensive Security - Total Defence." In *Strategic Cultures in the Asia-Pacific Region*, edited by Ken Booth and Russel Trood, 247–73. London, UK: Palgrave Macmillan.

Han, Eng Juan. 1996. "Preparing to Fight the Digital War." *Asian Defence Journal*, February, 20.

Ho, Peter. 2003. "Transforming MINDEF and the SAF." *Ethos - Singapore Civil Service College*, June, 1–5.

———. 2006. "Opening Address by Mr Peter Ho, Head, Civil Cervice at the Administrative Service Dinner & Promotion Ceremony." *Singapore Public Service Division*, March 30.

Ho, Shu Huang. 2009. *The Hegemony of an Idea: The Sources of the SAF's Fascination With Technology and the Revolution in Military Affairs*. IRASEC Discussion Paper No. 5, Singapore: IRASEC, 1–22.

———. 2015. "Total Defence Against Threat of Hybrid Warfare." *The Straits Times*, May 12.

Huxley, Tim. 1993. *The Political Role of the Singapore Armed Forces' Officer Corps: Toward A Military-Administrative State?* SDSC Working Paper No. 279. Canberra, Australia : Strategic and Defence Studies Centre, Australian National University.

———. 2000. *Defending the Lion City: The Armed Forces of Singapore*. St Leonards, Australia: Allen & Unwin.

International Institute for Strategic Studies. 1967–1975. *The Military Balance 1967, 1970, 1975*. London, UK: IISS.

Keling, Mohamad, Shukri Shuib, and Mohd Ajis. 2009. "The Impact of Singapore's Military Development on Malaysia's Security." *Journal of Politics and Law* 2 (2): 68–79.

Khoo, Jimmy. 2004. "Keynote Address by Future Systems Architect, BG Jimmy Khoo at C4I Asia Conference." Ministry of Defense, February 23, 1–9.

Koh Swee Lean, Collin. 2012. "Seeking Balance: Force Projection, Confidence Building, and the Republic of Singapore Navy." *Naval War College Review* 65 (1): 75–91.

———. 2014. "'Best Little Navy in Southeast Asia': The Case of the Republic of Singapore Navy." In *Small Navies: Strategy and Policy for Small Navies in War and Peace*, edited by Ian Speller, Deborah Sanders, and Michael Mulqueen, 117–31. Burlington, VT: Ashgate Publishing.

Kong, Kam Yean. 2007. "The Impetus for SAF Soldier Modernization." *Military Technology*, September 10: 87–88.

Lee, Edwin. 2008. *Singapore: The Unexpected Nation*. Singapore: ISEAS Publishing.

Lee, Hsien Loong. 1984a. "Speech at the National University of Singapore Forum on Scientist, 9 January 1982." *ASEAN Forecast* 4 (10): 164.

———. 1984b. "Security Options for Small States." *The Straits Times*, November 6, 1.

Lee, Jacqueline, Melvyn Ong, Ravinder Singh, *et al*. 2003. *Realising Integrated Knowledge-Based Command and Control: Transforming the SAF*. Singapore: SAFTI Military Institute.

Lee, Kuan Yew. 1998. *The Singapore Story: Memoirs of Lee Kuan Yew.* Singapore: Times Publishing Group.

——. 2000. *From Third World to First: The Singapore Story 1965–2000.* Singapore: Times Publishing Group.

——. 2009. "The Fundamentals of Singapore's Foreign Policy: Then & Now." *S. Rajaratnam Lecture at MFA Diplomatic Academy*, April 9, 5.

Leifer, Michael. 2000. *Singapore's Foreign Policy.* London, UK: Routledge.

Leong Sek Kay, Michael. 1995. "Application of Advanced Military Technology in Desert Storm." *Pointer – Journal of Singapore Armed Forces* 21 (1).

Lim, Chuan Poh. 2003. "IKC2: Transforming the SAF in the Information Age - A Foreword by Chief of Defence Force." In *Realising Integrated Knowledge-Based Command and Control: Transforming the SAF*, edited by Jacqueline Lee, Melvyn Ong, Ravinder Singh, *et al.*, 5–11. Singapore: SAFTI Military Institute.

Lim, Kim Choon. 2004. "Interview With Maj Gen Lim Kim Choon, Chief of Air Force, RSAF." *Asian Defence Journal*, January 10, 14–16.

Lim, Seng Hock. 1998. "Revolution in Military Affairs and Command, Control, Communications and Computers (C4): Are We Ready?" *Pointer – Journal of Singapore Armed Forces* 24 (2).

Loo Fook Weng, Bernard. 2004. "Explaining Changes in Singapore's Military Doctrine: Material and Ideational Perspectives." In *Asia in the New Millennium*, edited by Amitav Acharya and Lee Lai To, 352–74. Singapore: Marshall Cavendish Academic.

——. 2005. "Transforming the Strategic Landscape of Southeast Asia." *Contemporary Southeast Asia* 27 (3): 388–405.

——. 2015. "The Management of Military Change: The Case of the Singapore Armed Forces." In *Security, Strategy and Military Change in the 21st Century: Cross-regional Perspectives*, edited by Jo Inge Bekkevold, Ian Bowers, and Michael Raska, 70–89. London, UK: Routledge.

Low, Linda. 2002. "The Limits of a City-State: Or Are There?" In *Singapore in the New Millennium: Challenges Facing the City-State*, edited by Derek Da Cunha, 1–26. Singapore: ISEAS Publishing.

Matthews, Ron, and Nellie Zhang Yan. 2007. "Small Country 'Total Defence': A Case Study of Singapore." *Defense Studies* 7 (3): 376–95.

Menon, R. 1995. *To Command: The SAFTI Military Institute.* Singapore: Landmark Books.

Ministry of Defense Singapore. 1995. *Defense of Singapore 1994–95.* Singapore: Ministry of Defense.

——. 2000. *Defending Singapore in the 21st Century.* Singapore: Ministry of Defense Singapore.

——. 2007. *40/40: 40 Years & 40 Stories of National Service.* Singapore: Landmark Books.

Ng, Eng Hen. 2012. "Reply by Minister for Defence Dr Ng Eng Hen to Parliamentary Question on Singapore's Declining Birth Rate and Impact to the Singapore Armed Forces." Ministry of Defense, November 12.

——. 2015. "Speech by Dr Ng Eng Hen, Minister for Defence, at Committee of Supply Debate 2015." Ministry of Defense, March 6, 1–9.

Ng, Pak Shun. 2005. *From 'Poisonous Shrimp' to 'Porcupine': An Analysis of Singapore's Defence Posture Change in the Early 1980s.* SDSC Working Papers No. 397. Canberra, Australia: Strategic and Defence Studies Centre, 1–58.

Ong, Weichong. 2011. "Peripheral to Norm? Peripheral to Norm? The Expeditionary Role of the Third Generation Singapore Armed Forces." *Defence Studies* 11 (3): 541–58.

Raska, Michael. 2007. "Soldier-Scholars' and Pragmatic Professionalism: The Case of Civil-Military Relations in Singapore." APISA Occasional Paper Series No. 12.

——. 2014. "Submarine Modernisation in East Asia: Competitive Strategies." *RUSI Newsbrief* 34 (5): 24–26.

Raska, Michael, and Richard Bitzinger. 2015. "Russia's Concept of Hybrid Wars: Implications for Small States." *RSIS Commentary*, April 14.

Republic of Singapore Air Force. 2008. *40 Years of the RSAF: Our People, Our Air Force.* Singapore: RSAF.

Republic of Singapore Navy. 2007. *Onwards and Upwards – Celebrating 40 Years of the Navy.* Singapore: SNP International.

Singh, Bilveer. 2003. *Arming the Singapore Armed Forces: Trends and Implications.* Canberra, Australia: Strategic and Defence Studies Centre.

Tan, Andrew. 1998. *Singapore's Defence Policy in the New Millennium.* SDSC Centre Working Paper No. 322. Canberra, Australia : Strategic and Defence Studies Centre, Australian National University, 1–18.

——. 2001. *Domestic Determinants of Singapore's Security Policy.* APCSS Occasional Paper Series. Honolulu, HI: Asia-Pacific Center for Security Studies, 1–23.

Tan, Tony. 2002. "Interview with Dr Tony Tan, Deputy Prime Minister and Minister of Defence, Republic of Singapore." *Asian Defence Journal*, March 10, 10–11.

Tay, Gek Peng. 1999. "Joint Vision 2010: The Concept of Future Warfighting for the US Armed Forces and its Relevance to the SAF." *Pointer – Journal of Singapore Armed Forces* 25 (2).

Tay, Ronnie. 2007. "Interview with Rear Admiral Ronnie Tay - Chief of Navy, Republic of Singapore Navy." *Asian Defence Journal*, May 10, 14–21.

Teo, Chee Hean. 2004. "Statement at the 2004 Budget Debate in the Singapore Parliament." *Parliamentary Debates Official Report - Tenth Parliament*, March 15.

——. 2008. "Speech by Mr Teo Chee Hean, Minister for Defense, at DSTA-DSO Scholarship Award Ceremony." Ministry of Defense Singapore, July 8.

——. 2010a. "Speech by Deputy Prime Minister and Minister for Defence Teo Chee Hean at the Committee of Supply Debate." Ministry of Defense, March 5.

——. 2010b. "Meeting the Challenges of Singapore Defense." *Military Technology* (February): 16–18.

——. 2011. "Speech by Deputy Prime Minister and Minister for Defence Teo Chee Hean at the Committee of Supply Debate 2011." Ministry of Defense, March 2, 1–8.

The Straits Times. 2014. "Singapore Population Now at 5.47 Million, Slowest Growth in 10 Years." September 25.

6 Military innovation paths and patterns of small states

Comparative assessments

Previous chapters have traced the conceptual history and diffusion of military innovation in the context of changing security strategies of Israel, Singapore, and South Korea over the past two decades. Evidently, the three states have historically faced an array of persistent and complex security challenges brought by the realities of their external and internal security conditions and asymmetries of their location, size, and geopolitical constraints. With the end of the Cold War and subsequent shifts in the international and regional strategic environment, however, the key drivers, sources, and characteristics shaping their security environment have experienced profound changes. Their security environment has become increasingly uncertain and volatile with the emergence of "hybrid" security threats, which combine conventional, asymmetrical, low-intensity, and nonlinear threat dimensions. With the changing strategic realities, Israel, Singapore, and South Korea have been searching for a new security paradigm at the strategic level with concomitant operational adaptation and capabilities that would allow greater flexibility, adaptability, and autonomy under conditions of strategic uncertainty. In doing so, their militaries have explored, benchmarked, debated, and tested the applicability of select RMA-oriented concepts while attempting to acquire, develop, and integrate next generation of select weapons platforms, systems, and technologies as force multipliers of their existing military capabilities. The development of Israeli, Singaporean, and South Korean military innovation paths and patterns can be thus framed in the broader context of their national security debates on how to respond to the new security dilemmas – both in the current and long-term perspectives.

In a retrospective, however, their military innovation paths indicate predominantly *evolutionary trajectories, albeit with varying patterns*: In the process of continuous operational adaptation, the IDF has remained largely an establishment focused on "techno-tactical innovation" and "battlefield improvization" that avoids a "disruptive paradigm shift." In South Korea, given the single predominant threat from North Korea, the ROKA has been entrenched in its traditional conventional doctrine and institutional conservatism augmented by the U.S.–ROK alliance considerations. Consequently, there has not been a distinct Korean RMA-oriented conceptual innovation toward a new theory of war. In the case of Singapore, the process of 3G SAF transformation has reflected a structured-phased adaptation whether in terms of concepts, organizations, technologies, and capability development. At the same time, however, all three states have been able to mitigate, if not reverse, the asymmetries of their baseline geostrategic inferiority factors. Indeed, the IDF, ROKA, and SAF have, with varying levels of success, implemented organizational and technological adaptations that have amplified their military effectiveness vis-à-vis their existing or potential adversaries by pursuing a form of *niche* military innovation. In Israel, for example, a focus on immediate full-spectrum threats has facilitated "solution-oriented innovation" that has aimed to significantly shorten the traditional "sensor-to-shooter" cycles. In doing so, Israel's operational experimentation has shaped select military innovation paths and patterns of great powers. Meanwhile, Singapore has been able to offset an array of potential security predicaments through a comprehensive defense framework embedded in civil–military strategic interactions of the policy of "Total Defense," which has transcended purely military-technological domains.

This book argues that the varying trajectories result from the confluence of three sets of variables: (1) the level of strategic, organizational, and operational adaptability in responding to shifts in the geostrategic and regional security environment; (2) the ability to identify, anticipate, exploit, and sustain *niche* military innovation – select conceptual, organizational, and technological innovation intended to enhance the military's ability to prepare for, fight, and win wars; and (3) strategic culture. Specifically, strategic adaptability can be viewed as the ability to change defense strategy and military posture rapidly and seamlessly amid changes in geostrategic or regional security environment, while operational adaptability refers to the flexibility and robustness of armed forces to employ strategies and tactics in different ways and scenarios (Davis 2000, 31). Strategic culture can be defined as a function of the magnitude, character, and impact of "strategic path dependence" – embedded in the continuity and change of prevailing traditional security paradigms that set conditions for the development of strategic and operational thought within armed forces. Through the dynamic interaction of these variables, military innovation and its diffusion, particularly among small states, is not a hierarchical or linear process propelled by technological competition among great powers and then selectively adopted or emulated by less advanced "peripheral" or smaller units. Rather, it can take multiple facets and rarely proceed in a synchronized rate, path, or pattern. Given the prevailing external and internal variables – enablers and constraints that shape the capacity of states to integrate, adapt, and utilize *niche* military innovation under local circumstances – the process is not sequential nor does it follow a particular model. Technological innovation may precede conceptual and organizational adaptation, or conceptual speculation may lead to exploration and experimentation, but not technological implementation. Only if military innovation meets implementation in both policy and strategy, one can theorize about ideal "disruptive" defense transformation. Seen from this perspective, this chapter aims to assess the underlying factors shaping the varying military innovation trajectories in Israel, Singapore, and South Korea over the past two decades.

Frameworks assessing military innovation

Any comparative study must first define relevant indicators, performance metrics, and select frameworks for analysis. A closer look on the vast theoretical literature and debates in strategic studies concerning military innovation indicates the need for a comprehensive approach, integrating both theory and policy. Indeed, if national security planners are to make appropriate strategic decisions, they need to understand not only key theoretical debates but also relative trajectories of military innovation from a policy-oriented comparative perspective. To begin with, it is important to recognize the value of methodological pluralism and options in selecting both qualitative and quantitative methods that may yield different approaches and frameworks for analysis. While quantitative models, econometrics, and data processing may enable measurement of the rate of one or more innovations throughout the system, qualitative models are more useful in providing insights into the sources, conditions, enablers, and constraints that shape organizational capacity to identify, adapt, and exploit military innovation. In other words, a qualitative comparative approach may provide a detailed analysis on the varying transmission models, patterns, drivers, and levels of military innovation – that is, from duplicative and creative imitation or emulation to more sophisticated patterns of creative adaptation, architectural, and modular innovation to disruptive innovation. Furthermore, a qualitative comparative design may provide a distinct lens for identifying key policy challenges, problems, and implications in adopting and adapting select innovative concepts and technologies in defense planning and management processes in select states. Such analysis may then generate policy-oriented findings with regard to the pace, character, and success or failure in adopting and adapting military innovation. Seen from this perspective, this section introduces two qualitative frameworks for a comparative assessment of military innovation trajectories in the select case studies: (1) *Modified RMA Diagnostics Framework* and (2) *RMA Diffusion Dynamics Framework*. Both frameworks essentially apply the methodology of "structured focused comparison," in which a uniform set of select variables are embedded into carefully designed case studies, providing a specific research map, metrics, and indicators for a comparison. Both frameworks also synthesize the most important dimensions and factors that shape the pace, scope, and character of its adoption and adaptation of military innovation. At the same time, both frameworks reflect parsimonious as well as rigorous qualities and are therefore seen as the most relevant for this study.

Modified RMA diagnostics framework

Adopting and adapting new approaches to warfare takes considerable time, particularly in the presence of historical, strategic, institutional, and sociocultural factors that may facilitate as well as constrain military innovation. One way to understand the underlying strategic contexts, processes, and outcomes in a comparative perspective is to combine select elements from the structure and research design of the "RMA Diffusion Diagnostics" model by Emily Goldman and "Patterns of Military

Innovation" by Thomas Mahnken (Goldman and Mahnken 2004, 1–21). The resulting framework has three main levels of analysis: The first category – "Motives and Models" – attempts to map the prevailing motives, drivers, responses, and aspirations for adopting and adapting novel technologies, ideas, and practices. In other words, why are the selected states adopting and adapting particular military innovation strategies, concepts, or technologies? Based on Goldman and Mahnken's initial "diffusion diagnostics" model, there are at least four overlapping sets of drivers propelling the diffusion and adaptation of military innovation: (1) security, (2) economy, (3) technology, and (4) institutions. Specifically, security-related drivers emphasize the emergence of new strategic and operational challenges, existing geostrategic and demographic predicaments, alliance obligations, security dilemmas, and military strategic competition. In this context, according to Emily Goldman, military organizations can be considered as both comparative and competitive institutions that closely monitor each other and calibrate their performance in relation to other militaries (Goldman 2004, 1). External competitive pressure may provide strong incentive for states to emulate the military practices of the most successful states in the system. A military capability proven successfully on the battlefield can stimulate responses abroad – to emulate, offset, or innovate. In contrast, emerging operational concepts and innovative technologies can be also transmitted through collaborative processes, norms, and practices within alliance frameworks. "Lessons learned" from recent operational experiences are often filtered through strategic cultures of organizations attempting to adopt, adapt, or respond to the innovations elsewhere. Meanwhile, economic drivers focus on the military–industrial complexes and defense–industrial innovation processes shaping defense policies and military practices to adopt select innovative operational concepts and technologies. Technological drivers highlight the comparative advantages of specific commercial–civilian technologies and their application into military domains. Last but not least, institutional drivers underscore the role of bureaucratic interpretations as well as norms, forms, and practices pursued abroad. Inherently, the range of triggers, motives, and drivers will likely vary in different strategic settings, organizational contexts, and stages of pursuing military innovation. Consequently, Thomas Mahnken links the various "driver categories" into two main domains: "Opportunities and Threats." He argues that select states and military organizations are pursuing new approaches to combat in order to address existing security predicaments that defy conventional solutions. At the same time, they are attempting to exploit the benefits and advantages of emerging innovative operational concepts, organizations, and technologies. These resulting opportunities include the advantages of allied interoperability, expanded operational envelope, the introduction of new capabilities as force multipliers, and attaining comparative military advantage vis-à-vis strategic competitors (Mahnken 2004, 218).

Having mapped the varying motives and models of military innovation as opportunities and threats, the next level of analysis focuses on the assessment of *Paths and Patterns* of military innovation – assessing the varying stages of military innovation process: speculation, experimentation, and implementation (Mahnken 1999, 26–54). According to Thomas Mahnken, the first phase, *speculation*, involves identifying novel ways for solving existing operational problems or acknowledging the potential of emerging technologies. Its key indicators include the emergence of academic and professional publications – books, journal articles, concept papers, studies, authoritative statements, and speeches acknowledging and exploring the potential of new combat methods; establishment of official organizations or groups to study the lessons of recent wars; studies of foreign military innovation efforts; identifying intelligence collection requirements targeting country-specific innovation processes; informal debates focusing on identifying relevant military innovations and their relevance to country's defense strategies. As speculation turns into greater awareness, military services establish experimental organizations, battle laboratories, and units tasked with experimenting with new concepts, force structures, weapons technologies, and warfare methods. In the *experimentation phase*, military services form experimental units or innovative combat formations, which conduct training exercises specifically to test the applicability, utility, relevance, and potential effectiveness of new operational concepts, doctrines, or weapons systems. Wargaming, joint training, and other experimental activities reflect clear indicators of sustained interest in new warfare methods and technologies. In this stage, military services also attempt to interpret, evaluate, and assess the results of the experimental programs. Subsequently, the broadening and deepening experimentation processes reach a tipping point – a consensus – when the military leadership and services decide to adopt, adapt, and later refine select experimental operational concepts, warfare methods, organizational force structures, or new generations of weapons systems and technologies. The *implementation phase* may include a range of indicators: the establishment of new military formations; doctrinal revision to accommodate new ways of war; resource allocation supporting new concepts; development of formal transformation strategy; establishment of innovative military units; new branches and career paths; and, ultimately, field training exercises with new doctrine, organizations, or technologies (Mahnken 1999, 26–54).

The final "building block" in the analysis aims at mapping the varying *Barriers to Military Innovation* – the range of factors that shape (for better or worse) the capacity for military technology (hardware) and doctrine, tactics, and organizational structures (software) to be absorbed in different strategic environments. Emily Goldman suggests five main categories for analysis: polity, economy, society and culture, and military. In particular, an assessment of state's political system and environment, including factors such as state structure, governance, legal and regulatory frameworks, and civil–military relations, captures the state's capacity and preferences in mobilizing resources, converging institutional norms and interests, and fostering conditions to pursue innovation. Similarly, economic factors such as economic growth, industrial and technological resources, and defense spending play a role by either accelerating or hindering military innovation processes. There are also important sociocultural factors that include social structure, human capital, cultural resonance and tolerance, and organizational culture – together, these may accelerate or hinder the pace, character, and direction of military innovation (see Figure 6.1).

Level of Analysis		Category	Indicators
MOTIVES AND MODELS	**Threats, Challenges & Opportunities**	**Security**	Strategic Competition; Spheres of Influence; Alliance Obligations;
		Political Economy	Economic Pressures: by the Military-Industrial Complex; National Defense Community; Commercial Sector;
		Technology	Technological innovation driven by commercial incentives that encourage or discourage its adoption;
		Institutions	Bureaucratic/Organizational Competition; Institutional Norms, Forms, Practices for Emulation;
PATHS AND PATTERNS	**Sustaining & Disruptive**	**Speculation**	Conceptual Phase: Concepts, Books, Articles, Studies of New Combat Methods; Requirements Definition;
		Experimentation	Establishment of Experimental Organizations; Field Training, Exercises, and War gaming; Experimentation with New Combat Methods;
		Implementation	Formal Transformation Strategy; Revised Doctrine; Revised Military Education; New Practices that Refine Concepts;
ENABLERS AND BARRIERS	**Internal & External**	**Polity**	State Structure and Power; Political Diversity & Norms; Legal and Regulatory Framework; Security Focus of Armed Forces; Civil-Military Relations;
		Defense Economy	Weak economic growth; Limited defense resources; Limited defense industrial base; Dependence on external material support; Segregated defense sectors; Commercial focus; Export controls; Industrial and Technological Capabilities; Defense Spending;
		Strategic Culture	Social Structure; Human Capital; Cultural Resonance &Tolerance; National Culture;
		Organizational Force Structure	Inter-service rivalry; Asymmetry in power among services; Weak experimental base; Conflicts with innovation; Politicized military; Organizational Preferences; Weak Experimental Base; International Vulnerability; Organizational Type & Beliefs; Inter-connectness;

Figure 6.1 Modified RMA diagnostics framework

Source: Adapted from Goldman (2004) and Mahnken (1999).

RMA diffusion dynamics framework

Building upon the foundations of the above framework is the "RMA Diffusion Dynamics" model, conceptualized by the author of this study. It starts with the assumption that military innovation includes both internal and external processes of military emulation, adaptation, and innovation – whether in technologies, organizations, or doctrines. In other words, military innovation may not always require simultaneous technological, doctrinal, and organizational breakthroughs, but may span in the spectrum between incremental modernization and discontinuous transformation. Accordingly, the framework attempts to synthesize military innovation through the integration of: (1) *Conceptual Paths* – emulation, adaptation, and innovation; (2) *Technological Patterns* – speculation, experimentation, and implementation; (3) *Magnitude of Organizational Change* – exploration, modernization, and transformation. Through the confluence of both the "software" and "hardware" components in each domain, one can theoretically locate and visualize the diffusion of military innovation in a comparative perspective (see Figure 6.2).

In terms of conceptual paths, military emulation involves importing new ways of war through imitation of other military organizations. Adaptation is defined through adjustments of existing doctrines and methods, in which multiple adaptations over time may lead to innovation. Conceptual innovation then involves developing novel tactics, strategies, and structures. In the words of Theo Farrell and Terry Terriff, "it is only when these new military means and methods result in new organizational goals, strategies, and structures that innovation, adaptation, and emulation lead to major military change" (2002, 6). Applying Thomas Mahnken's framework, technological innovation may proceed in three distinct but often

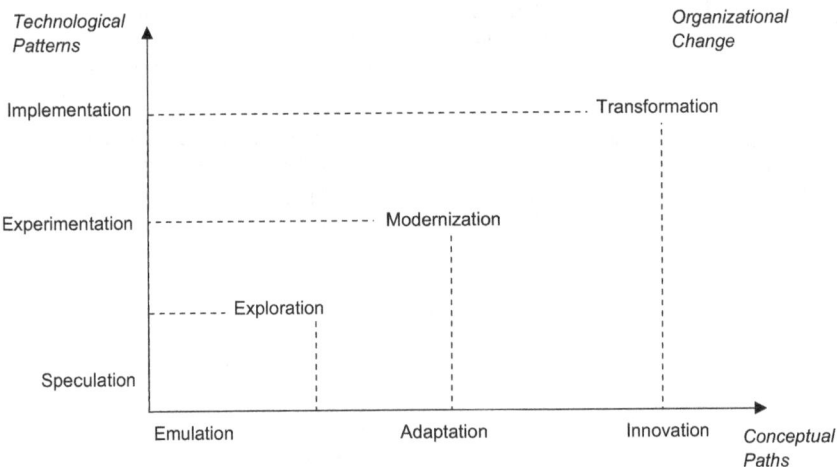

Figure 6.2 RMA diffusion dynamics framework

Source: Adapted from Mahnken (1999), Farrell and Terriff (2002), and Ross (2010).

overlapping phases: (1) speculation, (2) experimentation, and (3) implementation (1999, 26–54). By triangulating conceptual paths and technological patterns, it is possible to ascertain the character and magnitude of potential organizational change: exploration, modernization, and transformation (Ross 2010, 14–17). Exploration includes both speculation and emulation, with initial attempts to develop new areas of technological expertise; military modernization involves continuous upgrades or improvements of existing military capabilities through the acquisition of new imported or indigenously developed weapons systems and supporting assets (Tellis and Wills 2006). Transformation can be then characterized in the context of a "discontinuous" or "disruptive" military innovation that meets both policy and strategy. In the words of Andrew Ross, "disruptive, revolutionary innovation is the result of the confluence of discontinuous technological, doctrinal, and organizational changes; it occurs when discontinuous hardware and architectural changes coalesce and come together in a coherent, integrated whole. Existing capabilities are not optimized but rendered obsolete and displaced. New dominant technologies, doctrines, and organizations are established and integrated as never before. New performance metrics are embraced" (2010, 15).

Ultimately, the key measure of military innovation is in its output – in the process of developing, integrating, and maintaining select military capabilities in the operational conduct of armed forces. Military capability can be broadly defined in two categories: *as an output of national power* – the ability of a military force to successfully prosecute a variety of operations against a country's adversaries – and *as input measures* provided to the military and its ability to transform these resources into effective warfighting capability (Tellis, Bially, and Layne 2000, 133). In this context, military capabilities are neither constant nor absolute; they evolve (not always for the better) along with military organizations, training, equipment, and doctrine. Therefore, military capabilities are also difficult to measure: "the problems of measuring military capability are, in many respects, quite similar to the difficulties faced in measuring national power … any consideration of a country's military capabilities or its military effectiveness must begin with an examination of the resources – financial, human, physical, and technological – that the national leadership makes available to its military organizations" (Tellis, Bially, and Layne 2000, 135). Notwithstanding these inherent analytical challenges, one can still conceptualize the output of military capabilities based on two interrelated metrics: (1) *Systemic* – defined by a qualitative portfolio of available assets – advanced weapons technologies, systems, and platforms coupled with innovative operational concepts designed to perform new missions or significantly improve old systems or processes to achieve national security objectives; (2) *Functional* – defined by a portfolio of "capability domains" or defense competencies and functional tasks (i.e., battlespace awareness, command and control, net-centric operations, multi-mission air control, precision strike, layered defenses, stealth, etc.) in a particular set of operational missions and force deployments such as air, land, naval, and space operations (Davis 2002). In this perspective, an RMA-oriented innovative capability would enable a military organization to operate at higher levels of operational complexity (see Figure 6.3).

Capability (Functional) Domains			Operational Complexity
Ground Warfare	Air Power	Naval Warfare	
Knowledge-based warfare Adaptive Warfare Joint Warfare	Information Dominance Suppression of critical mobile targets Advanced suppression of enemy air defenses	Sea Control Multi-mission air control Deep strike	High
Full Combined Arms Coordinated Deep Attack	Offensive counter-air Advanced strategic strike Fixed-wing close air support	Naval Strike Limited air control Anti-submarine warfare	Mid
Basic Combined Arms Elementary Combined Arms Coordinated infantry Irregular Infantry	Basic Interdiction Basic defensive counter-air Elementary strategic strike Airspace sovereignty defense	Anti-surface warfare Anti-air warfare Coastal anti-surface warfare Coastal defense mining	Low

Figure 6.3 Assessing military capabilities

Source: Adapted from Davis (2002).

For example, select military units would be able to operate in (1) *depth* – having the ability to penetrate deep into adversary's territory and hit select targets with great precision; (2) *altitude* – having the ability to conduct both long- and short-range real-time surveillance and all-weather target acquisition and intelligence; and (3) *intensity* – having the ability to wage intense fighting (for a limited duration) integrating a range of air, ground, and naval platforms. Combining both systemic and functional approaches to military capabilities in a broader context of the Diffusion Dynamic framework may provide a comparative snapshot of the direction, character, trajectory, and magnitude of military innovation (see Figure 6.4).

Israel: from multiple operational adaptations to innovation

Since the mid-1990s, the IDF has pursued at least five publically known force modernization plans, both short term and long term: *IDF/Crossword* (2000), *Idan* (2003), *Idan* (2010), *Kela* (2008), and *Tefen* (2012). These have been periodically revised according to the changes in Israel's strategic outlook, defense needs, available budget levels, time horizons, and "lessons learned." In a broader perspective, however, the plans have reflected Israel's responses to the (1) continuously changing dynamics in Israel's security environment; (2) the budgetary pressures coupled with rising defense costs and the procurement of advanced weapons; (3) internal debates on the need to rethink IDF combat doctrines; and (4) strategic imperatives to enhance the IDF capabilities in order to maintain a qualitative strategic edge

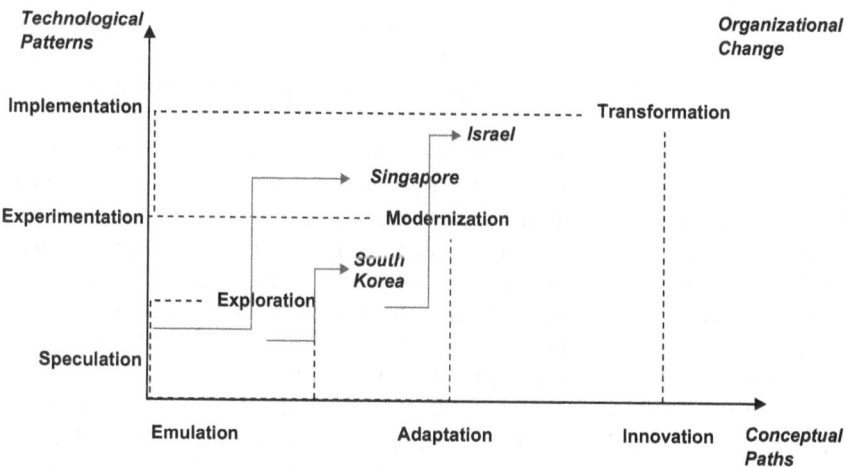

Figure 6.4 Comparing military innovation trajectories

Source: Adapted from Mahnken (1999), Farrell and Terriff (2002), and Ross (2010).

amid progressively complex security threats. Each plan also reflected a broader strategic imperative to transform the IDF toward "smaller but smarter" force. This concept, persistent in the IDF's discourse since the late 1980s, has entailed two interrelated aims: (1) streamlining IDF's force and command structure while leveraging advanced RMA-oriented technologies, and (2) make the IDF more professional and proficient (smarter) by upgrading the quality of the officer corps (Cohen 2008, 85). In particular, through specific defense reforms, the IDF would gradually mitigate its reliance on older platforms and force structures while assimilating advanced weapons technologies coupled with innovative operational and tactical concepts designed to enhance the IDF's military effectiveness at different levels of warfare. During the 1990s, the IDF selectively adjusted its existing concepts of "jointness" – combined operations by ground forces, navy, and air force; information superiority – networking of all intelligence assets and C4ISR capabilities to shorten the "sensor-to-shooter" cycle; and precision firepower. In the 2000s, the IDF experimented with an entirely new set of operational concepts based on the Systemic Operational Design (SOD) and attempted to translate them into specific processes and capabilities. These concepts, however, were seen as a natural extension or evolution in the conceptual innovation within the IDF. Nevertheless, they generated heated debates between its proponents and skeptics and subsequently shaped the character and direction of Israel's defense planning and management. In sum, underlying the various force modernization plans has been the imperative to retain Israel's military superiority amid increasingly complex high-threat environment and defense resource (budgetary) constraints. The formulation of each plan has been shaped by divergent strategic perceptions, contending schools of thought and debates within the IDF on the future direction in warfare, required resource allocation, and strategic and operational needs concomitant with the concept of "smaller and smarter" IDF. Arguably, it is within the context of these debates that the conceptual adoption and adaptation of the Israeli RMA-oriented strategic thought gradually gained momentum, particularly from the late 1990s onward.

Arguably, however, Israel has been the first country to use RMA-related advanced weapons technologies – that is, remotely piloted vehicles, integrated electronic warfare and real-time command, control, communications and intelligence systems, and precision munitions in combat in the 1982 War in Lebanon. Under the conceptual umbrella of "integrated battle," the IDF at that time focused on developing innovative techno-tactical solutions designed to mitigate threats and deficiencies, primarily those exposed during the 1973 Yom Kippur War. While these have not fundamentally altered IDF's concepts of operations or force structure, they have inherently shaped the direction and character of American, Soviet, and Israeli strategic perspectives and debates on the future of warfare (Adamsky 2010). Indeed, the Soviets were particularly alarmed by the military lessons of wars between Israel and its Soviet-armed Arab neighbors and began to reevaluate their strategic assessments. In doing so, they explored the conceptual contours of the emerging Military-Technical Revolution, which they viewed as bringing about a fundamental discontinuity in warfare with potential decisive implications for the Soviet Union. In the early 1990s, the U.S. intelligence community identified, monitored, and analyzed changes in the Soviet strategic thought, gradually acknowledged its significance, and subsequently developed theoretical underpinnings of the American RMA. In this period, select Israeli defense experts also began to acknowledge and speculate about the emerging notion of the RMA under the rubric of "future battlefield." The Israelis analyzed lessons from the U.S. military operations in the Gulf War, observing the combat potential and implications of new generation of military technologies. Notwithstanding the initial scepticism within the IDF of the significance of the U.S. military experiences during the first Gulf War, "Israelis [in the mid-1990s] have begun to recognize that there is indeed a revolution in progress" (Cohen, Bacevich, and Eisenstadt 1998, 92). Even at that time, however, the IDF has been at the forefront in employing select RMA weapons technologies and systems, albeit without acknowledging their "disruptive" character in changing Israel's security paradigm. Martin van Creveld notes that "already in 1991, when 80% of the ordnance dropped by the Americans on Iraq still consisted of 'dumb' iron bombs, the Israeli Air Force no longer had a single such weapons in its inventory; at that time the only UAVs in the U.S. arsenal had been made in Israel" (2005, 84).

The real impetus that sparked Israeli RMA-oriented conceptual drive, however, came with the establishment of the Operational Theory Research Institute (OTRI), which began to critically rethink nearly all aspects of IDF's traditional decision-making processes, assessment methods, and operational planning. Indeed, OTRI became an experimental laboratory for organizational and conceptual transformation within the IDF. Converging general systems theory with operational art, OTRI focused on the SOD, which formed a baseline for devising new methodologies for operational assessments for conducting warfare in nonlinear environments. During the ensuing decade, OTRI's educational courses, workshops, and ideas permeated into the IDF's senior ranks and regional commands, then to the General Staff, and subsequently were debated throughout the IDF. OTRI's influence culminated in the mid-2000s, when the IDF adopted new combat doctrine called "Concept of Operations" (CONOP), which combined select elements of network-centric warfare, precision stand-off firepower, and the SOD theory. Three months after its initiation, as select IDF units experimented and trained with CONOP's entirely new terminology, tactics, methods, and procedures, Israel became involved in the Second Lebanon War. In the one-month-long air and ground campaign, the IDF struggled to achieve its operational objectives vis-à-vis Hezbollah. After the war, its limited military performance and effectiveness was questioned, with many critics pointing to the complexity of CONOP as one of the key factors for IDF's operational shortcomings. Its proponents, however, argued that CONOP was not really implemented in the Second Lebanon War and has been widely misunderstood. With the follow-on assessments and lessons learned, the IDF initiated a comprehensive review of the CONOP doctrine while addressing IDF's organizational, structural, and operational deficiencies prevalent during the war. Subsequently, the IDF has discarded CONOP; however, it has not ceased searching for innovative operational and tactical solutions oriented toward particular threats, as seen in three rounds of conflict in Gaza: Operation Cast Lead (2008/09), Operation Pillar of Defense (2012), and Operation Protective Edge (2014). More recently, the IDF has increasingly integrated its *kinetic, intelligence, and cyber capabilities*, credited, for example, in the deployment of the *Stuxnet* code that disabled nuclear centrifuges in Natanz in 2010.

Taken together, Israel's "relational/action-oriented" pattern of operational adaption, speculation, and experimentation should also be seen in the context of Israel's tradition of military innovation and its four key drivers: (1) the basic, permanent strategic factors that traditionally characterized Israel's geostrategic circumstances (i.e., lack of strategic depth, regional demographic inferiority, resource constraints, etc.) and provided a primary impetus for Israel's military-technological priorities and constant focus on security (*ein brera* – no choice); (2) the historical development of Israel's indigenous military-industrial infrastructure responsive to the IDF's demands and capable of designing and redesigning a range of "combat-proven" weapons platforms, systems, and technologies; (3) the diverse operational requirements, needs, and experiences that have shaped the IDF's warfighting capabilities; (4) the culture of organizational learning and peculiar qualities of Israeli soldiers who reflected specific traits of IDF's strategic culture: resilience, creative improvisation under conditions of strategic uncertainties and prevailing resource constraints, flexible and pragmatic operational planning, egalitarian and self-reliant ethos, forward leadership and initiative, informal decision making, "to the point" communication, performance-oriented problem solving, demanding exercise training, sense of practicality, sensitivity to casualties, and learning from military failures

(Cohen, Bacevich, and Eisenstadt 1998). The nexus of strategic necessity, operational requirements, organizational culture, and technological innovation has propelled Israel to pursue new approaches to combat and seek advantages of new weapons technologies. While the process of acknowledging limits, failures, and formulating key "lessons learned" has never been "smooth," given the constant frictions, internal divisions, and debates, the critical security discourse has centered on relatively shared conceptions of threats and problems (see Figure 6.5).

Israel's RMA path over the last three decades thus reflects a unique pattern of early adoption/implementation, experimentation, and speculation in the context of multiple operational adaptations, which subsequently led to technological/tactical military innovation. As shown in previous chapters, the development of Israeli RMA concepts and technologies should be conceptualized primarily through the lens of continuity and change in Israel's security conceptions, focused on retaining Israel's "strategic edge" amid continuously changing operational requirements. The changing strategic realities over the last two decades essentially forced the IDF to rethink its traditional threat-oriented concepts and experiment with innovative capabilities-oriented concepts. In the process, however, the IDF has not initially viewed the emergence of the RMA as a disruptive paradigm shift, nor has initiated a comprehensive defense transformation drive. Rather, the IDF has gradually mitigated its reliance on older platforms and force structures, while assimilating advanced weapons technologies into existing force structures, and experimenting with selected innovative operational and tactical concepts at different levels of warfare. In doing so, the IDF has aimed to create conditions of "reverse asymmetry" at both the strategic and operational levels by creatively exploiting the IDF's qualitative superiority in both technology and human capital in order to offset its relative quantitative force inferiority (see Figure 6.6).

In the process, however, the IDF has also faced broader array of complex challenges, limitations, and constraints emanating from the confluence of political, economic, military, and sociocultural factors that have mitigated the effectiveness of Israel's defense/military innovation and, perhaps more importantly, its use of force. To begin with, the IDF had to cope with managing the complexities of military activities at two interrelated dimensions – *vertical* and *horizontal*. The vertical dimension includes the realities of preparation and conduct of war at the political, strategic, operational, and tactical levels (Millet, Murray, and Watman 1986, 2). In Israel's case, the IDF had to increasingly prepare and adapt for three types of military commitments simultaneously: (1) terrorism, guerrilla, and low-intensity warfare; (2) conventional operations; and (3) long- and short-range ballistic missile threats coupled with WMD threats. These threats were reflected in Israel's three circles of defense: perimeter, intra-frontier, and remote circle. During the Cold War, the predominant focus was on the "perimeter circle" that defined the frontlines of superpower rivalry in the Middle East and stipulated major conventional threats relevant to Israel's basic security. In contrast, since the early 1990s, Israel's strategic outlook has been shifting to a mixture of intra-frontier and remote military commitments

	Strategic Necessity	Operational Requirements	Strategic Culture	Technological Innovation
Threats & Challenges	– Geopolitical predicaments: location, geography, size; demography; – Proximity to areas of conflict; – Superpower involvement; – Historical Experience	– Multi-level Conflicts: – Conventional high-intensity war; – Low-intensity conflict; – Asymmetric conflict; – Non-linear conflict; – Expanded range of threats;	– Strategic uncertainty and pessimism–Siege mentality; – Low-casualty tolerance; – Focus on the objective; – Small margin of error;	– Budgetary constraints and cutbacks; – Technological risks – i.e. overreliance; – Asymmetric responses by potential opponents; – Narrowing technological gaps;
Opportunities	– Early Warning; – Deterrence; – Defense; – Preemption/ Prevention – Rapid military decision;	– Multi-mission capable forces; – Rapid sensor-to-shooter cycle; – Precision engagement; – Extended stand-off; – Force protection; – Increased combat power; – Reduction of force structure;	– Forward leadership & initiative; – Pragmatism; – Creativity; – Improvisation/ Flexibility – Organizational learning; – Egalitarian culture; – Emphasis on quality, austerity, and simplicity;	– Indigenous defense-industrial innovation – keeping technological edge; – Civil-military spin on/off effects;

Figure 6.5 Israel's military innovation: drivers, motives, and models

	Concepts	Patterns	Paths
1980s	"Integrated Battle"	**EARLY ADOPTION/IMPLEMENTATION** – Establishment of innovative combattactics; – Field training exercises with new weapons technologies; – Allocation of resources to support new warfare methods;	**CONTINUOUS MULTIPLE OPERATIONAL ADAPTATIONS** ↓ **MILITARY INNOVATION**
1990s	"Future Battlefield"	**SPECULATION** – Publications describing potential new combat methods; – Establishment of official organizations to study recent wars; – Study of foreign innovation efforts;	
2000s	"Systemic-Operational Design"/ CONOP	**EXPERIMENTATION** – Experiments with new methods of warfare; – Establishment of innovative combattactics – Allocation of resources to support new warfare methods;	
2010s	"Kinetic-Cyber" Operations	**IMPLEMENTATION** – Establishment of innovative combattactics; – Allocation of resources to support new warfare methods; – Operational implementation;	

Figure 6.6 Israel's military innovation: concepts, paths, and patterns

concomitant with non-linear, asymmetric forms of warfare, spanning between "sub-conventional" and "super-conventional" threats (Cohen and Inbar 1991). In this context, the boundaries between the types of conflict and the actors involved have blurred, forcing the IDF to reassess traditional threat perceptions and adapt to complex multifaceted missions under severe operational tempo requirements. Subsequently, the nearly constant focus, if not overload of, "current security issues" resulted in short-term orientation and intellectual aversion that precluded long-term conceptual, organizational innovation, or comprehensive defense transformation (Adamsky 2010). Moreover, with the widening range of military commitments, IDF's operational conduct and performance has been increasingly questioned and scrutinized, both internally and externally (Creveld 1998). While much of the criticism can be tracked, the problems and misconceptions in Israel's fragmented political decision making coupled with changes in Israel's civil–military relations – that is, problems in defining clear, attainable political and strategic objectives in the diverse IDF campaigns (i.e., Second Lebanon War in 2006) – a substantial part of the critiques focused on the IDF's military capabilities, force structure, operational concepts, and fighting standards (Kober 2008).

The strategic and operational challenges constraining an Israeli RMA in the vertical dimension have inherently also transcended into the *horizontal* – pertaining to the numerous, simultaneous, and interdependent defense management tasks, requirements, and processes to prepare for multilevel conflicts. These include defense planning, strategic analyses and assessments, defense budgeting, training, logistics, weapons procurement, intelligence, technical adaptation, risk management, and related functions aimed to maintain specific capabilities and levels of military readiness. With the changing character of conflicts coupled with persisting resource limitations, Israeli defense planners have faced complex challenges in defense management. First, in terms of strategic and intelligence assessments, the process of identifying the probability and magnitude of potential threats in relation to costs and performance of military requirements has stipulated new levels of organizational complexity. While 30 years ago the IDF focused predominantly on a single "circle" of defense and conventional warfighting capabilities, the variables shaping the post-Cold War strategic assessments have progressively widened and the margins of error in the evaluations have narrowed. Moreover, the traditional organizational boundaries of responsibility between the various intelligence branches (i.e., the Military Intelligence AMAN, the Mossad, and the General Security Service Shin Bet) have become more intertwined (Eiland 2007). Furthermore, while Israel has been relatively successful in developing and sustaining the intellectual and technological capabilities to meet indigenous needs, it has faced significant budgetary challenges. In particular, as the performance of advanced military technologies increased in relation to their costs (and vice versa), the task of allocating resources and setting priorities between the three IDF services became more competitive in terms of balancing conflicting interests and managing trade-offs. The Israeli Navy, for example, has trailed behind IDF's air and land components in procurement priorities and often struggled to find resources for complex development programs and follow-on production requirements. Select acquisition and procurements programs in other branches have also faced long delays in their operational integration amid various budgetary, bureaucratic, and technical problems (Opall-Rome 2008). Indeed, the IDF has for years faced constant budgetary pressures and increasing demands to find the most versatile and flexible solutions that would provide generic capabilities for a maximum number of conflict scenarios. Specifically, how to balance defense resources in order to maintain specific levels of military responsiveness (immediate deployment capabilities), operational readiness (short-term, monthly training, operating and maintenance costs), size of the armed forces (mid-term, multi-year force structure), and R&D and future systems procurement (long-term acquisitions)? These challenges have been debated by the Israeli security establishment for decades, particularly in the aftermath of each "round" of military conflict in and around Israel's circles of defense (see Figure 6.7).

	Policy	Planning	Management	Op. Conduct
Political	Competing definitions of national security;	Identifying the probability and magnitude of potential threats;	Identifying service-priorities and risks in procurement programs;	Blurring boundaries of organizational responsibilities;
Strategic	Strategic uncertainty and multi-level conflicts;	Costs and performance of military requirements;	Persisting Resource Constraints;	Short-term orientation; Intellectual aversion in strategic culture;
Operational	Widening military commitments;	Persisting Resource Constraints;	Technical/ Organizational problems in systems integration;	Leveraging new concepts and technologies in practice;
Tactical	Focus on "Current Security" issues;	Changing work plans; exercise plans; field operations plans;	Devising relevant training programs: military responsiveness vs. readiness;	Adaptation challenges to new roles, missions, and tactics;

Figure 6.7 Israel's military innovation: barriers and constraints

South Korea: from conceptual emulation to technological adaptation

For the past two decades, South Korea has been pursuing comprehensive defense reforms in order to respond to the widening spectrum of threats, mitigate techno-logical and interoperability gaps with the U.S. forces, and eventually attain self-reliant defense posture. South Korea's RMA-oriented military modernization drive has its roots in the early 1990s, when the ROK's Ministry of Defense, ROK Armed Forces, and a small number of Korean/U.S. defense analysts began to conceptual-ize long-term force modernization visions based on the evolving strategic priori-ties and increasing scope of defense requirements. With the accelerating RMA drive in the U.S. military, the ROK political and military echelons began to explore the changes in the dynamics of modern warfare, the implications of the RMA, and to devise new strategies for comprehensive defense reforms. Accordingly, one could argue that the emergence of the RMA paradigm in the United States had a signifi-cant impact on South Korea's perceptions of future warfare and concomitant defense management processes. South Korean defense planners studied the evolving U.S. RMA debates and attempted to adapt select concepts into their long-term defense plans and force improvement programs. In doing so, they initially developed a concept of "omnidirectional, advanced force," which became the cornerstone of the

Korean RMA drive in the late 1990s. While the concept did not represent a unique theoretical or conceptual innovation with regard to a new theory of war, it was part of a broader national security debate on the current status and potential future direction of the ROK forces as well as the U.S.–ROK alliance. From the early 2000s onward, the United States and South Korea embarked on a process of adapting their alliance to the changes in the U.S. global defense posture and reconfiguring the roles, missions, and existing command structures. The seemingly techno-oriented RMA conceptualization in the U.S. military shifted into a much broader and ambitious process of defense transformation. In 2005, the United States and South Korea agreed to transform the U.S. strategic presence and operational conduct in Korea. Under the revamped alliance system, the ROK forces would be required to increase their combat capabilities and assume greater autonomy and responsibility in defense of the country, while provide operational support to increasingly agile U.S. forces. At that time, South Korea's defense policy emphasized the so-called "cooperative self-reliant defense" – the need to mitigate South Korea's dependence on the United States by gradually developing indigenous or self-reliant military capabilities that would enable to deter any existing threats from North Korea while preparing for unspecified future threats. Underscoring the policy became the vision of transforming the ROK military into a smaller, but increasingly networked, balanced, and digitized standing force (the concept of "Digitized Defense") with independent surveillance and reconnaissance platforms, real-time integrated C4I systems covering a variety of tactical, operational, and strategic echelons, and long-range precision strike capabilities (ROK Ministry of National Defense 2004). The changing security dynamics in conjunction with the shifts in the U.S.–ROK alliance subsequently shaped the *Defense Reform Plan 2020* (*DRP 2020*) – arguably the most comprehensive, ambitious, transformation-oriented force modernization plan. Introduced in the fall of 2005, the plan was based on strategic assumptions that the North Korean threats would inherently diminish by 2020, while potential intra-regional threats coupled with regional force modernization drives may emanate new security challenges for South Korea. Through the *DRP 2020*, South Korea's forces would significantly enhance their military capabilities, expanding the scope and reach of their "operational horizons" (Bechtol 2006). Many new weapons platforms as well as their components and sub-systems would be based on Korea's indigenous R&D defense industrial base, with foreign sources associated with the supply of major items and leading edge technologies (Johan 2004). With its ambitious scope, proposed timeline, and high costs, however, the *DRP 2020* propelled internal policy debates on the feasibility, affordability, pace, direction, character, and implementation of the plan. In 2010, following the sinking of the South Korean corvette *CheonAn* and the artillery shelling of the Yeonpyeong Island, the direction of the *DRP 2020* and much of South Korea's defense planning has been completely overhauled. In March 2011, South Korea's Ministry of Defense announced *Defense Reform 307*, primarily addressing medium- to long-term defense requirements to counter potential North Korean asymmetric provocations, infiltrations, and attacks similar to the sinking of the *CheonAn* and artillery attack on the Yeonpyeong Island. However, similarly to previous plans, the 307 Plan has faced considerable debates and opposing views. Reportedly, selected high-ranking officers in the ROK Navy and the Air Force have opposed the plan amid concerns that the new integrated command structure may lead to an "Army-dominated joint force." Others have questioned the implementation feasibility of the plan itself amid previous *DRP 2020* budget shortfalls. The trajectory of South Korea's military modernization plans thus suggests a progressive complexity in South Korea's defense planning, a search for a relevant defense strategies, force posture, and operational concepts that would allow greater flexibility, adaptability, professionalism, and autonomy to address existing as well as future-oriented security challenges and defense requirements (see Figure 6.8).

Having a significant U.S. imprint, South Korea's RMA path can be conceptualized on two parallel levels: (1) *external* – shaped by the changes in the U.S. defense strategies, operational concepts, and force postures; (2) *internal* – embedded in South Korea's efforts to minimize prevailing U.S.–ROK interoperability gaps, respond to widening security challenges, and simultaneously develop more "self-reliant" military capabilities, particularly at the various inter-service levels. Arguably, the former has shaped the latter by providing key benchmarks for defense planning, interoperability requirements, and operational conduct of the ROK forces. Selected U.S RMA-oriented concepts, that is, information superiority, precision-strike, battlefield situational awareness, and network-centric warfare, have gradually permeated into the U.S.–ROK combined training and operations and subsequently shaped the character and direction of South Korea's RMA-oriented defense reforms, weapons acquisition and procurement, and conceptualization of future warfare. Accordingly, South Korea's RMA diffusion path – at the conceptual level has been relational, reflecting characteristics of *operational USFK emulation* and *internal adaptation*, with limited or no disruptive innovation. Indeed, while South Korea has increased its force development expenditures, and initiated a number of diverse acquisition and procurement programs of selected advanced weapons platforms and systems, the composition, structure, and deployment of the ROK forces have remained relatively unchanged since the Korean War. In other words, while the ROK's armed forces have acknowledged the strategic imperative for an RMA-oriented military modernization, it has not been able to transform the ROK military's organizational force structure, nor design innovative strategic and operational concepts on how conflicts should be analyzed and waged. The path dependence has been entrenched in the confluence of (1) political, historical, and strategic legacies, (2) persisting traditional threats amplified by the emergence of more complex and uncertain threats, (3) widening interoperability and technological gaps in the U.S.–ROK alliance, (4) budgetary constraints, (5) prevailing conservatism and politicization in South Korea's strategic culture, and (6) lack of diverse combat experiences have all precluded the implementation of South Korea's defense reforms (see Figure 6.9).

Seen from this perspective, South Korea's RMA-oriented military modernization shows patterns of speculation and experimentation in terms of concepts and military technologies, but with limited implementation in the corresponding changes of the organizational force structure. Indeed, one can argue that ROK forces have faced a number of political, strategic, operational, and tactical impediments, anchored in South Korea's traditional security paradigm, which

	Strategic Necessity	Operational Requirements	U.S.–ROK Alliance	Defense Management
Threats & Challenges	– Current Threats / Proximity to areas of conflict (North Korea); – Emerging geostrategic predicaments (Rise of China); – Unresolved historical legacies (Japan); – Regional force modernization;	– Current Threats (North Korea): – Uncertain Future Threats: Asymmetric warfare; Energy security; Cyberwarfare; Terrorism; Maritime Security;	– Strategic dependence on the U.S.; – Changes in the U.S.–ROK Alliance; – Shifts in the regional U.S. force posture; – Technological and interoperability gaps with U.S. forces;	– Budgetary constraints; – Technological risks and costs; – Political debates-fragmentation of the security consensus; – Operational maintenance of existing force; – Land-centric force structure; – Conservative strategic culture;
Opportunities	– Self-reliant defense, – Capabilities critical for emerging combat areas; – Early warning; – Deterrence; – Strategic adaptability;	– Multi-mission capable forces; – High-speed maneuverability – Precision targeting and strike; – Maritime and sea-control; – Airpower; – Balanced force structure and jointness;	– Interoperability with U.S. forces; – Operational War Control (OPCON) – Coalition warfare; – Enhanced operational readiness and responsiveness;	– R&D of weapons systems reflective of a Korean strategic environment; – Indigenous defense-industrial innovation;

Figure 6.8 South Korea's military innovation: drivers, motives, and models

	Concepts	Patterns	Paths
1990s	Air Force Vision 2030 Navy Vision 2020 Army Vision 2010	**SPECULATION** – Study of U.S. innovation efforts; – Publications describing potential new combat methods; – Establishment of official organizations to study recent wars;	**EMULATION**
2000s 2010s	Omni-Directional, Advanced, Elite Force Defense Reform 2020 Defense Reform 307	**EXPERIMENTATION** – Allocation of resources to support new warfare methods; – Field training exercises with new weapons technologies; – Capability development;	**CONCEPTUAL ADAPTATION** **LIMITED IMPLEMENTATION**

Figure 6.9 South Korea's military innovation: concepts, patterns, and paths

have precluded a major defense transformation and inhibited military innovation. To begin with, the ROK military in the post-Cold War era had to formulate new strategic and combat doctrines that take into account three seemingly contrasting requirements: (1) capabilities for a wider spectrum of threats emanating from North Korea; (2) devising selective capabilities for a post-North Korean threat-based military force, particularly in the context of accelerating regional military modernization drives, actions and strategies of Japan, China, and Russia; (3) preparing the U.S.–ROK military relationship, institutions, and mechanisms for an array of challenges linked to potential Korean unification scenarios – that is, civil–military challenges of occupation while combating a capable, well-armed insurgency. In this context, however, the evolving and diverse nature of inter-Korean relations over the past decade has polarized South Korea's political arena, with *persisting debates* on the magnitude and character of North Korean threats; terms and conditions of potential Korean unification and its implications; changes in U.S. strategy and levels of American security commitment to South Korea; and concomitant short and long-term strategic requirements, force posture, and defense resource allocation. Two broad camps and lines of thought have emerged in these debates: *softliners* – seeking peaceful coexistence with North Korea through cooperation and exchange, calling for a more reserved defense posture and the resolution of regional strategic instability through multilateral diplomacy; *hardliners* – advocating more assertive strategic planning, force structure, and joint U.S.–ROK responses to North Korean threats and provocations. Both camps have experienced their prevalent "highs and lows" under different administrations and their policies toward Pyongyang (i.e., "Sunshine Policy" of Kim Dae-Jung and Roh Moo-Huyn vs. "realist" policy of Lee Muyng-bak vs. Park Geun-hye's "Trustpolitik"). The popular support for both camps, however, shifted with North Korea's destabilizing moves, provocations, and crises that challenged prevailing policy assumptions and threat perceptions in the South, resulting in even wider internal divides and strategic uncertainties. In other words, the increasing fragmentation of South Korea's political arena has arguably led to an *erosion of strategic consensus*, which subsequently resulted in *contrasting calibrations* in South Korea's defense planning processes. Moreover, the political will to allocate the required defense resources for implementing selected defense reforms has been constrained by *economic pressures* and imperatives to sustain South Korea's socio-economic stability and growth. Indeed, South Korea has struggled to find finances to adequately fund the various military modernization plans, which resulted in delays in the procurement of selected advanced weapons platforms, systems, and technologies, particularly air force and naval modernization programs.

At the strategic level, after nearly six decades of largely *static, defensive posture* focused on defending the DMZ and *reliance on direct U.S. support*, the ROK military has been constrained by its own institutional rigidity, intellectual conservatism, and path dependence, embedded in the traditional force structure and deployment centered on conventional ground forces. According to Chung-in Moon, South Korea's existing force structure has been defective on three accounts: (1) it is *not adequately equipped* with dealing with both present and future threats arising from non-conventional and high-tech war scenarios in the region – the preoccupation with traditional conventional threats has fundamentally undermined defense readiness for newly emerging contingencies; (2) it has been *structurally overdependent on U.S. forces* in the areas of C4ISR as well as naval capabilities and air power – with concurrent risks arising from a potential abrupt reduction or withdrawal of American forces; and (3) it lacks credible second strike capabilities to ensure effective deterrence (Moon 2000). In this context, one could further add: (4) *traditional conscription* unable to provide more professional soldiers capable of conducting advanced joint operations with the air force and navy. The institutional resistance to change may have thus emanated from the traditional *inter-service rivalries* with the *ROKA-dominated leadership* unwilling to shift defense resources to the air and naval forces, and substantially alter the ROK's force structure. The Army's rationale, stemming from South Korea's traditional security paradigm, is that South Korea continues to confront a serious conventional threat from North Korea's robust offensive military capabilities, and in the absence of timely and accurate intelligence on the evolving political and military situation in North Korea, the ROK National Command Authority and adjacent organizations must be prepared for such contingencies (Lee 2003). Moreover, notwithstanding South Korea's qualitative superiority in the current conventional balance of forces, North Korea retains a substantial quantitative advantage in all aspects, which are amplified by asymmetric force multipliers – ballistic missiles and WMD programs. In this line of thought, South Korea's force structure must be centered on the primacy of ground forces supported by a robust U.S.–ROK Alliance, with the ROKA as the largest military service and primary responsibility for defending South Korea. However, this strategic logic became increasingly questioned over the last decade amid the gradual transformation in the character of North Korea's calibrated strategies, which became focused on asymmetric forms of warfare to mitigate the effectiveness of South Korea's deterrence capabilities and force improvements (see Figure 6.10).

Ultimately, the confluence of political, economic, and strategic constraints embedded in South Korea's traditional security paradigm precluded the implementation of selected defense reforms also at the operational level, where the ROK forces have been striving to overcome a range of technical and interoperability problems in conducting full-spectrum military operations, that is, *obtaining a common operational picture, coordinating intelligence, ensuring information security, enabling cooperative command, control, and communication capabilities, maintenance of existing systems, and field deployment and interoperability of new systems*. These challenges can be attributed to the traditional force structure design and its ramifications, that is, inter-service divides, technological and operational deficiencies with the U.S. forces, and arguably, *the lack of direct and diverse*

	Policy	Planning	Management	Op. Conduct
Political	Internal political fragmentation: i.e. NK, U.S.– ROK Alliance;	Contrasting calibrations of defense requirements;	Persisting resource constraints and limitations; i.e. budget cuts;	Inter-service rivalry; ROK Army dominated force structure;
Strategic	Structural dependence on the U.S.– ROK Alliance;	Costs and risks of required defense reforms;	Shifting priorities in selected procurement programs;	Structural dependence on the U.S.–ROK Alliance;
Operational	Widening military commitments; hybrid security threats;	Technological interoperability gaps with the U.S. forces;	Technical/ Maintenance Problems; Deployment of new weapons;	Reliance on U.S. forces: C4ISR, naval capabilities, air power;
Tactical	Static, defensive force posture; Traditional conscription;	Lack of credible second strike or retaliatory capabilities;	Devising joint training programs; systems integration;	Adaptation to asymmetric threats: devising new combat tactics; Lack of diverse combat experiences;

Figure 6.10 South Korea's military innovation: barriers and constraints

combat experience. Notwithstanding its limited participation in selected PKO missions (East Timor, Lebanon), non-combatant operations (Iraq and Afghanistan), and major joint/combined training and exercises in South Korea, including annual exercises such as the "Ulchi Focus Lens" (UFL), "Reception, Staging, Onward Movement and Integration" (RSOI), "Foal Eagle" (FE) "Amnokgang," and "Hoguk," South Korean forces have not gained credible warfighting experience since their participation in the Vietnam War. This does not mean that South Korea has not maintained a qualitative edge over North Korea's forces, but it shows the link between South Korea's long-standing, static, defensive posture emphasizing conflict and war avoidance and relative overreliance on the U.S. forces, with the resulting operational inertia to pursue a more disruptive military reforms.

Singapore: structured-phased technological adaptation

Since the early 2000s, Singapore has pursued its military innovation drive embedded in the concepts, technologies, and processes of the 3G SAF force transformation. Arguably, the primary driver for the SAF's military innovation has not changed – *strategic necessity*. First, the SAF's 3G concepts, processes, and debates proceeded along established security paradigms based on the confluence of both internal and external threat perceptions, historical legacies, geostrategic and demographic factors, considerable resource dependencies, and continuity and change in Singapore's defense requirements. These included Singapore's baseline geographical disposition factors – location, size, lack of natural resources, and physical limitations of Singapore as a small island city–state; lack of strategic depth as well as natural resources; close proximity to potential regional flashpoints and territorial disputes in the South China Sea; risks of maritime trade interdictions or transnational crimes at sea; and the prevailing asymmetries in demographic and population factors – whether internal or external. In this context, the SAF's primary mission has remained relatively constant: deter any threats to Singapore's security, territorial integrity, and sovereignty, enhance Singapore's peace and security (through defense diplomacy), and if deterrence or defense diplomacy fails, enable a swift and decisive victory. The confluence of these factors predicated a sustained robust resource allocation in the development of credible defense capabilities. Conceptually, these have evolved from the 1G SAF (1960s–70s) that aimed on basic capability development of individual services; the 2G SAF moving toward conventionally oriented combined arms warfare (1980s–90s), and the 3G SAF (2000s – onward) reflecting a period of experimentation and transition toward a joint inter-service strategy with multi-mission capabilities ranging from defense diplomacy to integrated strike against wide spectrum of threats. In this context, Singapore's strategic thought and doctrine has also evolved from a purely island-defensive "poisoned shrimp" strategy in the 1970s, which envisioned high-intensity urban combat to impose unacceptable human and material costs to potential aggressors, toward "porcupine" strategy of the 1980s that developed a limited power projection in Singapore's near seas and envisioned a preemptive posture by transferring a potential conflict into enemy's territory, to the ongoing 3G transformation in the 2010s analogous to a "dolphin" strategy – a "smart" or networked SAF leveraging not only precision fires, maneuver, and information superiority capabilities but also operations other than war in geographically more distant areas from Singapore.

At the same time, however, Singapore's military innovation drivers also include *external factors*: strategic shifts in East Asia's security landscape characterized by contending relations based on unresolved historical legacies coupled with emerging security challenges, tied primarily to China's rise and resulting regional superpower competition. Indeed, East Asia's strategic assessments and debates have centered on five key issues: the pace, character, and direction of China's military modernization; the struggle for dominance by the region's two major powers (China and Japan); the future of the Korean Peninsula; intra-regional competition in territorial disputes in the East China Sea and South China Sea; and perhaps most importantly, the contours of long-term regional strategic competition and rivalry between China and the United States. East Asia's security template has been therefore shifting toward a mix of asymmetric anti-access/area-denial threats, low-high intensity conventional conflicts in traditional flashpoints such as the Taiwan Strait or the Korean Peninsula, and a range of nontraditional security challenges such as energy security, cyber security, and intra-regional competition in territorial disputes in East China and South China Seas. Notwithstanding the increasing regional economic linkages and widening prosperity underscored by China's economic rise, most Southeast Asian countries have shared concerns about China's growing military capabilities and future aspirations in the region. Indeed, as a result of China's "coercive diplomacy" since mid-2010, select Southeast Asian states have been revamping their force modernization priorities, alliances, and overall strategic choices. These trends have not led to a regional "arms race" per se (i.e., action-reaction cycle of arms acquisitions based on mutually adversarial relationships, explicit tit-for-tat arms acquisitions, the intention of seeking dominance over one's rivals through arming and intimidation), but rather to a gradual "arms competition" or "arms dynamic" characterized by a mix of cooperative and competitive pressures, continued purchases of advanced weapon platforms, including the introduction of new types of arms and, therefore, unprecedented military capabilities (Bitzinger 2010).

Furthermore, Singapore's military innovation trajectory has been strongly driven by its continuously improving *defense management capacity*, embedded in the symbiotic relationships, links, and perhaps most importantly, a "unitary vision" between the key players in Singapore's "defense ecosystem" – that is, the users (SAF), developers (i.e., MINDEF, DSTA, DSO dual-use R&D labs), and producers (local defense industries, i.e., ST Engineering). In particular, Singapore's "cost-benefit" system of defense acquisitions, including its Research, Development, Technology, and Engineering domain, has traditionally reflected strong preferences for cost-effective, technological solutions adapted for Singapore's unique defense requirements. In doing so, close links and coordination among the Singapore Government, the SAF, and Singapore's defense-industrial base, exemplified in a competitive technocratic approach in

R&D investment, weapons acquisition and development processes, and experimentation, have enabled effective interoperability and accelerated assimilation or technological system integration across Services (Ho 2009).

Seen from this perspective, the SAF's military innovation trajectory can be characterized as a *structured-phased adaptation*; a strategy of capability development balancing current and future defense requirements along three track vectors: (1) *Force Readiness*, (2) *Force Evolution*, and (3) *Force Transformation* (Lim 2003). Force readiness implies constant organizational reshaping of the SAF that would preserve its operational readiness – the ability to rapidly transform from a disciplined and well-prepared force during peacetime to an effective fighting force during wartime, capable to conduct diverse missions in an expanded spectrum of operations. Force evolution has aimed at systematic development of existing capabilities based on the current conceptual and doctrinal frameworks. Force transformation has focused on creating capacity for a long-term conceptual, organizational, and capability transition toward a joint strategy based on continuous experimentation, organizational learning, and operational experience. It is in the latter that the SAF adopted its 3G transformational vision of multi-mission capable, technologically superior, and organizationally networked forces, effectively defending Singapore from current as well as emerging new threats. In this context, however, the concept of 3G SAF cannot be excluded from the continuity and change of Singapore's "Total Defense" strategy – a form of national security strategy that has for over 30 years aimed at strengthening and mobilizing resources in five mutually supportive defense domains: military, civil, economic, social, and psychological. In other words, Singapore's military innovation drive has been a part of a *comprehensive defense innovation framework* leveraging civil–military strategic interactions (see Figure 6.11).

Singapore's paths and patterns of military innovation thus project a classical trajectory of *speculation, experimentation, and implementation* – an ongoing process of strategic adaptation, organizational learning, capability development, and operational implementation. First, the SAF studied the emerging RMA debate in the United States since the mid-1990s and subsequently explored its applicability and impact on the SAF – a form of speculation. In the 2000s, the SAF accelerated its intellectual development and experimentation in the context of the 3G force transformation by developing the conceptual, organizational, and operational aspects of Integrated Knowledge-based Command and Control (IKC2) across the Services – a form of experimentation. The IKC2 has taken its cue from the U.S. strategic thought, initiatives, and operational experience with "Network-centric Warfare" and "Effects-based Operations." It envisioned the networking of sensors and forces by extending the information domain and creating conditions for the emergence of network-centric operations. In other words, IKC2 implied a symbiotic relationship between shared information, increased shared situational awareness, and increased combat power. At the same time, IKC2 focused on creating environment that supports organizational learning, leadership, and innovation, particularly in the operational adaptation of networking of the varying "sensors and shooters" of the SAF. The task for disruptive-type experimentation, whether

	Strategic Necessity	Operational Requirements	Disruptive Technologies
Threats & Challenges	– Baseline geographical disposition factors – location, size, lack of natural resources, lack of strategic depth; – Historical experience; – Perennial sense of vulnerability;	– Close proximity to potential regional flashpoints and territorial disputes in South China Sea; – Widening operational requirements: from operations other than war to conventional deterrence;	– Global diffusion and advances in Information and Communications Technologies; – Changing character of combat: asymmetric warfare;
Opportunities	*Force Readiness:* Capabilities to operate in expanded spectrum of operations;	*Force Evolution:* Systematic development of existing capabilities based on the current conceptual and doctrinal frameworks;	*Force Transformation:* Long-term conceptual, organizational, and capability transition based on continuous experimentation, organizational learning, and operational experience;

Figure 6.11 Singapore's military innovation: drivers, motives, and models

conceptual, organizational, or technological, was led primarily by the Future Systems Directorate and Defense Research and Technology Office, which cooperated with the SAF Services as well as Singapore's defense R&D institutions in developing and testing select advanced technologies, systems, and concepts in relations to new ways of warfare. From the late 2000s onward, the SAF focused on the implementation of the 3G SAF vision. The implementation phase can be also characterized as a structured-phased or building-block approach: first, the SAF acquired new equipment and introduced progressively more capable systems coupled with the establishment of new units to enable the transformation of the SAF into an advanced, networked force; second, the SAF established new operational commands to deal with an expanded spectrum of operations, at home and overseas, and in doing so, focused on widening its operational flexibility and responsiveness. And in third phase, the SAF aimed on enhancing its leadership and human capital through the introduction of enhanced career schemes as well as revision of training and curriculum, to maintain a steady stream of capable and committed people to meet the requirements of the 3G SAF (Teo 2010). From 2015 onward, the SAF began to explore emerging operational concepts and strategies concomitant with "hybrid warfare" that combine asymmetric, non-linear threats with select elements conventional warfare. Hybrid warfare is perceived as a future strategic challenge to the future of Singapore's national defense; at the same time, however, Singaporean defense planners may find ways to use hybrid warfare to its own advantage. Singapore has developed a highly capable intelligence apparatus, which can detect and counter influence operations; it has a strong sense of patriotism and "total defence" education, which reinforces national resiliency and can help citizens resist foreign propaganda and other types of compulsion and a professional air and naval force, which can function as a screening force to give the army time to mobilize an adequate homeland defense. At the time of this writing, however, Singaporean defense community is studying as to how hybrid warfare might specifically impact Singaporean defense, from where it might emanate, and how it might manifest itself. The initial assumption is that SAF's future "kinetic strike capabilities" would have to be complemented by greater societal cohesion and resilience, seen as fundamental to the social, economic, and psychological pillars of Total Defense (see Figure 6.12).

While the SAF's technological, organizational, operational, and professional capabilities continue to qualitatively outpace its neighbors in relative terms, it also faces a number of challenges, limitations, and barriers – particularly in pursuing military innovation. To begin with, throughout its history, the SAF in conjunction with Singapore's "defense ecosystem" has developed the ability to identify, assimilate, integrate, and manage advanced defense technologies to enhance the operational capabilities of the SAF. In doing so, the principal focus has been on achieving cost effectiveness and self-sufficiency, gradually replacing the reliance on overseas suppliers for select high-end military technologies, while maintaining rigorous institutional oversight in technical evaluations, systems-of-systems engineering, and robust government–military–industrial collaboration. In a process that could be characterized as administrative, technocratic, and cost-effective defense

	Concepts	Patterns	Paths
1990s	"RMA Future Battlefield Studies"	**SPECULATION** – SAF allocates resources to study recent wars; and supports new warfare methods;	**CONTINUOUS STRUCTURED-PHASED ADAPTATION**
2000s	SAF 3G Force Transformation Block -1 IKC2 – "Integrated Knowledge Command and Control"	**EXPERIMENTATION** – SAF publications describing potential new combat methods; – Establishment of SAF Future Systems Directorate – to study foreign innovation efforts; and emerging or disruptive technologies; – Acquisition of progressively more capable systems coupled with the establishment of new units;	
2010s	SAF 3G Force Transformation Block - 2 & 3	**IMPLEMENTATION** – SAF sets up new operational commands to enable expanded spectrum of operations; – Leadership and human capital development; – Revision of training and curriculum;	
2015-	"Hybrid Conflicts" "SAF Force 2030"	**SPECULATION / EXPERIMENTATION** – SAF allocates resources to study emerging advanced technologies, operational concepts;	

Figure 6.12 Singapore's military innovation: concepts, paths, and patterns

management, Singapore's military modernization has evolved through clearly predefined phases bounded by institutional parameters and rigid "top-down" oversight. On one hand, this process has enabled the development of professional competencies, but at the same time, discouraged multidisciplinary "bottom-up" alternative approaches. Singapore's defense ecosystem has thus evolved into an adaptive "systems integrator" rather than a "disruptive innovator."

This leads to the question of continuity of challenges posed by the authors of *Creating Capacity to Change* more than a decade ago: *how to nurture a hotbed of truly innovative ideas alongside regular, well-functioning bureaucracy?*

How do we strike the right balance between central control, which ensures uniformity of standards and clear direction for the different parts of our organisation, and decentralised freedom, which gives individuals sufficient space to innovate and unleash their potential? In which areas should we strive for self-sufficiency, as opposed to reaping the efficiencies of outsourcing? How can we achieve inter-temporal optimisation of resources between training for current readiness and re-invention for future capability? When should we respect the accumulated wisdom of established procedures, and when should we adopt new practices that are more relevant to new circumstances?

(Choy *et al.* 2003, 12)

Indeed, one could argue that the above questions are still relevant for the SAF, perhaps more than ever. While the SAF has succeeded in organizational learning and tactical problem solving, the development of its long-term strategic thought has remained in unitary control and unquestionable authority by the senior leadership. Singapore's tightly controlled political and socioeconomic environment has thus imposed preconceived ideas and solutions through deeply rooted hierarchical structures, processes, and narratives, which have effectively discouraged the emergence of "intellectual outliers" nurtured in a "culture of creativity." Singapore's perennial "existential angst" has led to risk aversion if not risk avoidance, both in the civil and military domains. In the SAF, its strategic culture of administrative day-to-day operational management has de facto precluded an environment supporting long-term "maverick" visions that would challenge the established schools of thought, paths, and patterns, particularly at the strategic level. Increasingly, the SAF has also demonstrated greater technological dependencies, given limited manpower resources, which led to excellent tactical experimentation at the operational level but, at the same time, invoked growing perceptions of "technological superiority" as the primary strategic determinant, enabler, and catalyst of military effectiveness. Notwithstanding these cultural/organizational challenges, Singapore's model of structured-phased adaptation has reflected incremental, often near-continuous, improvements of existing capabilities, augmented by the select introduction of advanced weapon platforms, which have gradually enhanced Singapore's military capabilities – from conventional warfare to operations other than war. With the increasing complexity of East Asian security dilemmas, particularly with respect to the convergence of "old" and "new" security threats such as deepening territorial disputes and potential crises over islands in the South China sea, SAF's operational requirements will likely increase further. Accordingly, the SAF will have to strike a balance between preserving tried-and-tested strategies and structures with finding innovative operational concepts and organizational structures in preparation for multilevel conflicts. The key dilemma facing Singaporean defence planners will remain constant: *how to build a force and doctrine capable of dealing simultaneously with current security threats while anticipating future challenges.*

References

Adamsky, Dima. 2010. *The Culture of Military Innovation: The Impact of Cultural Factors on the Revolution in Military Affairs in Russia, the US, and Israel.* Palo Alto, CA: Stanford University Press.

Bechtol, Bruce. 2006. "Force Restructuring in the ROK-US Military Alliance: Challenges and Implications." *International Journal of Korean Studies* X (2): 19–41.

Bitzinger, Richard. 2010. "A New Arms Race? Explaining Recent Southeast Asian Military Acquisitions." *Contemporary Southeast Asia* 32 (1): 50–69.

Choy, Dawen, Ju-Hon Kwek, Chung Han Lai, *et al.* 2003. *Creating the Capacity to Change: Defence Entrepreneurship for the 21st Century.* Singapore: SAFTI Military Institute.

Cohen, Eliot, Andrew Bacevich, and Michael Eisenstadt. 1998. *Knives, Tanks, and Missiles: Israel's Security Revolution.* Washington, DC: Washington Institute for Near East Policy.

Cohen, Stuart. 2008. *Israel and Its Army: From Cohesion to Confusion.* London, UK: Routledge.

Cohen, Stuart, and Efraim Inbar. 1991. "A Taxonomy of Israel's Use of Military Force." *Comparative Strategy* 10 (2): 121–38.

Creveld, Martin. 1998. *The Sword and the Olive: A Critical History of the Israeli Defense Forces.* New York, NY: Public Affairs.

———. 2005. *Defending Israel: A Controversial Plan Toward Peace.* New York, NY: St. Martin's Press.

Davis, Paul. 2000. "Defense Planning in an Era of Uncertainty: East Asian Issues." In *Emerging Threats, Force Structures, and the Role of Air Power in Korea,* edited by Chung-in Moon and Natalie Crawford, 25–47. Santa Monica, CA: RAND.

———. 2002. *Analytic Architecture for Capabilities-Based Planning, Mission-System Analysis, and Transformation.* Santa Monica, CA: RAND.

Eiland, Giora. 2007. "The Changing Nature of War: Six Challenges." *INSS Strategic Assessment* 10 (1): 1–10.

Farrell, Theo, and Terry Terriff. 2002. *The Sources of Military Change: Culture, Politics, Technology.* Boulder, CO: Lynne Rienner Publishers.

Goldman, Emily. 2004. "Introduction: Military Diffusion and Transformation." In *The Information Revolution in Military Affairs in Asia,* edited by Emily Goldman and Thomas Mahnken, 1–23. London, UK: Palgrave Macmillan.

Ho, Shu Huang. 2009. *The Hegemony of an Idea: The Sources of the SAF's Fascination with Technology and the Revolution in Military Affairs.* IRASEC Discussion Paper No. 5, Singapore: IRASEC, 1–22.

Johan, Saad. 2004. "ROK Armed Forces: Force Development Continues Unabated." *Asian Defence Journal* (10): 22–26.

Kober, Avi. 2008. "The Israel Defense Forces in the Second Lebanon War: Why the Poor Performance?" *Journal of Strategic Studies* 31 (1): 3–40.

Lee, Chung Min. 2003. "East Asia's Awakening from Strategic Hibernation and the Role of Air Power." *Korean Journal of Defense Analysis* 15 (1): 219–74.

Lim, Chuan Poh. 2003. "IKC2: Transforming the SAF in the Information Age – A Foreword by Chief of Defence Force." In *Realising Integrated Knowledge-Based Command and Control: Transforming the SAF,* edited by Jacqueline Lee, Melvyn Ong, Ravinder Singh, *et al.,* 5–11. Singapore: SAFTI Military Institute.

Mahnken, Thomas. 1999. "Uncovering Foreign Military Innovation." *Journal of Strategic Studies* 22 (4): 26–54.

———. 2004. "Conclusion: The Diffusion of the Emerging Revolution in Military Affairs in Asia – A Preliminary Assessment." In *The Information Revolution in Military Affairs in Asia*, edited by Emily Goldman and Thomas Mahnken, 211–24. London, UK: Palgrave Macmillan.

Millet, Alan, Williamson Murray, and Kenneth Watman. 1986. "The Effectiveness of Military Organizations." *International Security* 11 (1): 37–71.

Moon, Chung-in. 2000. "Changing Threat Environment, Force Structure, and Defense Planning." In *Emerging Threats, Force Structure, and the Role of Air Power in Korea*, edited by Natalie Crawford and Chung-in Moon, 89–114. Santa Monica, CA: RAND.

Opall-Rome, Barbara. 2008. "Debut of Israeli Trophy Delayed." *Defense News*, September 15.

ROK Ministry of National Defense. 2004. *Defense White Paper 2004*. Seoul, Korea: Ministry of National Defense.

Ross, Andrew. 2010. "On Military Innovation: Toward an Analytical Framework." *IGCC* (Policy Brief No. 1): 1–4.

Tellis, Ashley, Janice Bially, and Christopher Layne. 2000. *Measuring National Power in the Postindustrial Age*. Santa Monica, CA: RAND.

Tellis, Ashley, and Michael Wills. 2006. *Strategic Asia 2005–06: Military Modernization in an Era of Uncertainty*. Seattle, WA: The National Bureau of Asian Research.

Teo, Chee Hean. 2010. "Speech by Deputy Prime Minister and Minister for Defence Teo Chee Hean at the Committee of Supply Debate." *Ministry of Defense*, March 5.

7 Conclusion

Theoretical and policy implications

This book argued that military innovation paths and patterns in select small states have reflected an *evolutionary trajectory – a spectrum of military innovation* shaped by the confluence of three sets of interrelated factors: (1) the level of strategic, organizational, and operational adaptability in responding to shifts in their geostrategic and regional security environment; (2) the ability to identify, anticipate, exploit, and sustain *niche* military innovation – select conceptual, organizational, and technological innovation intended to enhance the military's ability to prepare for, fight, and win wars; and (3) strategic culture. In this context, the book showed the varying motives and drivers, enablers and constraints, and strategic and operational limitations in the military innovation paths and patterns of Israel, Singapore, and South Korea. The book argued that their military innovation trajectories must be first viewed in the context of their progressive complexity of security dilemmas, which have widened their strategic, organizational, and operational requirements in addressing both current as well as future-oriented security environment under the growing cloud of security uncertainties. In the process, the three states have selectively explored, benchmarked, and debated the applicability of emerging RMA-oriented schools of thought, while attempting to exploit the next generation of select weapon platforms, systems, and technologies. These technologies, however, have provided necessary but not a sufficient condition for the implementation of their military innovation aims. In order to respond to the changing strategic realities and security uncertainties over the past two decades, Israel, South Korea, and, to a lesser degree, Singapore recognized the need to consider smaller, professional, and multi-mission capable force structure backed by advanced military technologies that are adaptable, agile, and highly interoperable. At the same time, the three states have faced a number of complex political, economic, military, and organizational barriers, embedded in the path dependence of their traditional security paradigm, which have mitigated the pace and character of their force modernization and military innovation. In short, each state has reflected different levels of strategic and operational adaptability to military innovation – the receptivity to change military posture rapidly in response to shifts in the geostrategic conditions and having the flexibility to use new concepts and technologies in different military scenarios (Davis 2000, 31). In this context, it is important to extrapolate relevant theoretical implications (at the risk of oversimplification), which may help to draw conclusions for this study.

In Israel, the nexus of persisting security challenges coupled with operational and tactical experiences had arguably the most profound impact on the formulation and implementation of Israeli RMA-oriented military innovation, designed to sustain Israel's qualitative "strategic edge." Indeed, 30 years ago, the IDF has essentially pioneered today's network-centric warfare capabilities concomitant with stand-off precision strike, real-time battlefield intelligence, rapid target acquisition, integrated command and control, and remotely piloted vehicles. In doing so, however, the IDF has not viewed these innovations as disruptive or revolutionary, but rather as relevant adaptive measures that "fit new techniques, inventions, or operational outlooks into deeply rooted and relatively fixed military paradigm" (Cohen, Bacevich, and Eisenstadt 1998, 68). In time, however, the continuous adjustment in IDF's operational conduct led to multiple adaptations and bottom-up learning processes, which subsequently accelerated Israel's military innovation. As Dima Adamsky noted, "the technological seeds of Israeli RMA preceded the conceptual ones" (2010, 93). In short, Israel has been relatively more effective (albeit not always) in absorbing and exploiting select RMA-oriented innovative concepts and technologies – following multiple conceptual, technological, organizational, and operational adaptations at different levels of warfare over time – driven mainly by Israel's changing threat spectrum and subsequent search for operational and tactical solutions oriented toward particular threats.

In contrast, military innovation in South Korea shows patterns of speculation and experimentation in terms of concepts, doctrine, technologies, but relatively *limited implementation* in the integrated (joint) use of force – despite the strategic proximity and operational experience with U.S. forces. Indeed, for more than a decade, the ROK military has attempted to conceptualize and synchronize its own distinct "Korean RMA" by emulating and adapting select U.S. network-centric warfare concepts relevant for the Korean Peninsula. This parallel Korean RMA trajectory can be seen in South Korea's efforts to enhance its independent strategic planning, interoperability with U.S. forces, and, more importantly, its broader aim to achieve "self-reliant" military capabilities. Notwithstanding concerted efforts to create smaller but smarter "advanced elite force," Korean RMA drive has not fully eradicated the power of the "old" paradigm, which has been sustained by the confluence of factors, including contrasting calibrations of defense requirements, structural dependence on the U.S.–ROK alliance, and long-term static defensive force posture dominated by the ROK army. The lack of diverse operational combat experiences has also precluded greater flexibility and adaptability in translating select defense reforms in practice and propelled a more disruptive type of military innovation. Indeed, there has not been a distinct Korean RMA-oriented conceptual innovation toward a new theory of war. At the same time, however, with the changing character, magnitude, and impact of North Korea's asymmetric conventional and cyber security challenges, the ROK military has gradually pursued a military modernization drive, focusing on the procurement of select power projection platforms as force multipliers in combined air, land, sea, underwater, and cyber operations.

In Singapore, notwithstanding the absence of direct combat experiences, the SAF has been gradually pursuing military innovation under the conceptual and organizational umbrella of 3G force – a long-term process of transition toward a joint strategy from a platform-centric to network-centric multi-mission capable force. In an evolving structured-phased adaptation approach, embedded in a relatively constant security paradigm, the SAF has been focusing on adopting and adapting select technological force multipliers in order to offset Singapore's geostrategic limitations, regional security uncertainties, and to ensure its military-technological edge against potential future asymmetric and hybrid threats. In this context, the SAF's 3G implementation proceeded in three phases: In the first phase, the SAF acquired new equipment, introduced progressively more capable systems, and established new units to enable the transformation of the SAF into an advanced, networked force. In the second phase, the SAF has set up new operational commands relevant with an expanded spectrum of operations, at home and overseas, and, in doing so, focused on widening its operational flexibility and responsiveness. And in the third phase, the SAF aimed at enhancing its leadership development and organizational learning through the introduction of enhanced career schemes as well as revision of training and curriculum, to maintain a steady stream of capable and committed people to meet the requirements of the 3G SAF (Teo 2010). At the operational level, the SAF has been also experimenting with innovative concepts and tactics, particularly in the context of special operations, urban fighting, and operations-other-than-war under the conceptual umbrella of integrated knowledge-based command and control. Accordingly, Singapore's path and patterns of military innovation, therefore, reflect a classical trajectory of gradual conceptual speculation, organizational adaptation, and operational implementation. Equally important, however, has been Singapore's broader defense innovation transcending purely military-technological domains. Under the enduring concept of "Total Defense," Singapore's security has been embedded into five mutually reinforcing dimensions – military, civil, economic, social, and psychological security. Taken together, Singapore's strategy has pursued an integrated and adaptive defense posture based on twin-pillars of external deterrence and diplomacy, while strengthening its internal socioeconomic resilience and responsiveness against a range of potential non-traditional security threats and challenges. In the process of searching for military innovation, however, the SAF has also inherently faced a number of challenges: at the strategic level, the question of how to strike a balance between preserving tried-and-tested strategies and force structures (based on a traditional conscription model) with implementing advanced operational concepts, organizational structures, and technologies relevant for a new type of future "hybrid conflicts." At the operational level, the SAF has demonstrated greater technological dependencies that have, on one hand, propelled excellent "joint" tactical experimentation but, at the same time, invoked growing perceptions of "technological superiority" as the primary determinant, enabler and catalyst, of military effectiveness.

Based on above findings, no single theoretical bent can capture the dynamics of military innovation – the crucial interactions of both exogenous and endogenous factors that shape the inputs and outputs in military effectiveness. In particular, the *internal* factors such as human capital, leadership, financial, institutional, and organizational resources in the context of foreign and domestic policy making are all required to initiate and sustain the process of innovation – from speculation, experimentation, to implementation. The process of developing new ways of war in terms of technology, doctrine, and organizational changes often takes considerable time and does not follow a sequential order. Indeed, it is rare for military innovation to embrace both new hardware and software components simultaneously; new technologies may precede new concepts and processes, or vice versa. In this context, conceptual adoption and organizational adaptation often prove more challenging than the technological hardware acquisition. This is because the procurement and acquisition of new weapons systems and technologies requires rethinking of established principles and norms that guide the employment of military forces. At the same time, the effectiveness of military innovation is always relative to the *external* factors, primarily the adversarial responses in technologies, strategies, and aspirations aimed at offsetting the advantages of particular military innovation. Notwithstanding these inherent analytical complexities, there are at least three major contending schools of thought that predict whether a state is likely to adopt, adapt, and implement military innovation, and which may amplify the varying debates explaining the variance in this study: (1) neorealist (structural realist) perspectives, (2) organizational/societal perspectives, and (3) culturalist perspectives. Individually, each school is incomplete in relation to the other and bound to critical examination. By combining these theoretical perspectives, however, one may ascertain the three dimensions that shape military innovation trajectories – from strategy, organizational capacity, to culture.

Structural realist perspectives

Arguing that strategic competition, in the context of anarchic international relations, drives military innovation, neorealist (structural realist) perspectives predict that external security environment plays a primary role in shaping state's strategic choices. In particular, it is the insecurity – the presence of external threats – that provides key incentives for states to innovate. In this view, states with higher levels of insecurity will have stronger incentives to innovate. At the same time, states with expanding international interests and ambitions will also have stronger incentives to innovate in order to defend their interests, shape the international environment, and gain strategic advantage by power projection. In the process, the competitive nature of international relations will drive states to emulate the military capabilities and successful innovations, organizational forms, practices, and technologies of their rivals or superior powers in the system. As Kenneth Waltz noted, "the possibility that conflict will be conducted by force leads to competition in the arts and instruments of force. Competition produces a tendency toward the sameness of competitors" (1979, 128). Accordingly, military innovation in international system is uniform, hierarchal, with innovations appearing first in the most advanced states and then subsequently emulated and adapted by smaller powers in the way that fits their particular geopolitical circumstances, social structures, and available resources (Isaacson, Layne, and Arquilla 1999).

Neorealists would therefore explain the varying Israeli, Singaporean, and South Korean RMA trajectories in the context of their particular geostrategic settings that shaped their perceptions of insecurity and provided strategic rationale to pursue military innovation, that is, relative asymmetries of the location, geography, and size; natural resource constraints; dependence on external political and material support; security uncertainties or proximity to areas of conflict; and relations and importance between and to great powers. As small states from a relational point of view, Israel, Singapore, and South Korea have to varying degrees politically and strategically align with the United States, which provided political and material support including access, sales, and transfer of advanced weapons platforms, systems, and technologies that have been optimized to their select defense requirements. These technologies were subsequently adopted and adapted by the respective armed forces. The frequency and character of combat experiences coupled with innovative operational concepts and informal strategic culture, however, has enabled Israel to use them more effectively. As shown in Chapter 3, the IDF would take the initiative to conduct high-intensity precision strikes, transferring war into enemy territory, flanking and outmaneuvering its enemies to reduce casualties and achieve tactical surprise, while implementing operational and command flexibility that allowed more efficient adaptability to the changing circumstances. Meanwhile, South Korea's armed forces have focused on largely static, defensive posture, relying on the U.S. forces for deterring military threats in peacetime and defending South Korea in wartime. In this context, however, neorealist perspectives fail to explain Israel's RMA trajectory, characterized by "relational/action-oriented" pattern of multiple operational adaptations that led to technological and later conceptual military innovation, which has subsequently influenced the course of strategic thought as well as military technology development of great powers. Indeed, as far back as the 1973 Yom Kippur War, Israel's use of force drew the interest of both the U.S. and Soviet military and, in doing so, indirectly accelerated their technological as well as conceptual and doctrinal military innovation toward the RMA. In 1982, the United States adopted the ALB doctrine, while the Soviet Union began to intellectualize the contours of the emerging Military-Technical Revolution and its related Reconnaissance-Strike Complexes. In the early 1980s, Israel's creative use of remotely piloted vehicles and real-time C3I systems realistically anteceded both the U.S. and Soviet RMAs, with images of integrated or network-centric warfare, stand-off "smart" precision firepower, electronic warfare, and near-simultaneous engagement of multiple targets at greater distances (Adamsky 2010). For Israel, however, these have not represented disruptive transformation in warfare but rather a creative application of technotactical innovation and battlefield improvization under constantly changing operational conditions. It was until two decades later when the IDF's OTRI embarked for the first time on a comprehensive conceptual innovation drive centered on the theory and methodology of Systemic Operational Design, which was perceived as a major "lexical turn" in IDF's strategic and operational thought and later served as a baseline for the 2006 Concepts of Operations (CONOP) combat doctrine. While CONOP's operational applicability, relevance, and implementation has been contested and debated in the aftermath of the Second Lebanon War in 2006, eventually forcing the IDF to discard the doctrine, its select conceptual elements permeated into the U.S. military thought processes – that is, in the formulation of the U.S. Army's Campaign Design. Similarly, one could argue that Israel's experiences and techniques in low-intensity conflicts, counterterrorism, and asymmetric warfare have provided relevant insights for the U.S. military and its training for combat operations, particularly for the wars in Iraq and Afghanistan. In short, while the IDF has deployed select U.S. weapons technologies and has been familiar with U.S. operational concepts, its own unique operational experience and lessons learned – successes and failures – profoundly shaped the external theoretical development and understanding of the RMA. Consequently, one can argue that changes in strategy resulting from operational lessons learned in diverse combat experiences may increase the adaptability of military organizations to implement military innovation. Israel's combat experiences, its lessons learned from high- and low-intensity conflicts, have accelerated IDF's "bottom-up" military innovation by increasing the need to find practical solutions first rather than focus on theoretical conceptualizations.

On the other hand, South Korea has struggled to implement its comprehensive defense reforms amid the prevalence of the traditional security paradigm that has inherently precluded similar bottom-up military innovation. Notwithstanding its growing technological capabilities and, more importantly, its prolonged and close strategic and military proximity with the U.S. forces through the U.S.–ROK alliance, South Korea has been seemingly unable to implement U.S. concepts concomitant with network-centric warfare, nor translate the various comprehensive defense reforms. This represents a paradox given two additional variables: (1) the widely acknowledged strategic imperatives to enhance ROKA's military capabilities, and (2) the country's accelerating, broadening, and deepening information revolution synonymous with Korean IT innovation, digitization, and commercialization over the past decade, which has not profoundly translated into the military domain. Indeed, South Korea's existing operational concepts, force structure, and deployment have remained relatively constant for more than five decades. In order to improve operational effectiveness, interoperability with U.S. forces, and eventually attain defense "self-reliance," South Korea has focused primarily on the "hardware" side – allocating resource for the procurement and acquisition of select advanced weapons technologies, systems, and platforms, whether imported or indigenously developed, which were then integrated into *existing* organizational force structures and operational concepts. The required "software" component – the relevant organizational, conceptual, and doctrinal innovation to utilize the new technologies – has not been fully implemented.

In Singapore's case, its foreign and defense policy has been embedded in a consistent, conventional, realist perspective of a small state that has no choice but to acknowledge great powers in the international system, while coping with potentially hostile external environment. In doing so, however, Singapore has not pursued interlocking alliances, but rather a set of policies based on balance or distribution of power intended to contribute to regional stability and assure independence of the city-state. In this context, Singapore's military innovation drivers have aimed to develop and maintain cutting-edge military capabilities vis-à-vis potential opponents in the region, while ensuring that the city-state's economic interests are protected. In the process, Singapore has relied on the United States as the country's most important strategic partner, supplying about 60 percent of the country's imports in the past decade – primarily its airpower platforms. At the same time, however, Singapore has adapted technologies supplied by France, Germany, Italy, Sweden and Israel. Accordingly, in a realist perspective, one could conclude that Singapore has leveraged on international competition and partnerships to maximize technology transfers and, more importantly, develop indigenous capabilities to assimilate these technologies into its defense industrial base (DIB) as well as its armed forces. The 3G+ SAF's transformation has, therefore, not emulated a particular model of military innovation, but reflected a multitude of sources that have shaped Singapore's evolving paths and patterns trajectory of speculation, experimentation, and implementation with a strong emphasis on "self-reliance."

Organizational theory perspectives

Emphasizing the importance of organizational capacity to embrace and absorb technological changes, organizational and societal perspectives focuses on internal factors that may enable or constrain the implementation of military innovation (Isaacson, Layne, and Arquilla 1999). In particular, organizational theory perspectives focus on the role of civil–military relations in debating the question how and why military innovation occurs – whether by civil–military dynamics (i.e., the intervention of civilian policy makers into entrenched military organizations); inter-service competition between military branches; or intra-service competition for roles, missions, and resources (Posen 1984; Rosen 1991). In doing so, civil–military relations perspectives emphasize the level of societal cohesion in facilitating military effectiveness and innovation. As Stephen Peter Rosen theorized, "military organizations that are separated from their host society and which draw on that society for resources are in tension with that society. They extract resources while being different from and under representative of the larger society. This tension can and has created problems in prolonged war or prolonged peacetime competition. An innovative military that extracts resources but is isolated from society may not be able to sustain that innovation in periods of prolonged conflict" (1996, 1). In this perspective, innovation is less likely to occur in divisive societies, as the state tends to focus on internal problems and channels lesser resources to the military. On the other hand, cohesive societies tend to generate greater offensive and defensive power from available material resources. Similarly, intra-service institutional perspectives argue that military organizations are inherently conservative, entrenched, and resist innovation, as they are concerned primarily with internal distribution of status, gains, and power (i.e., in terms of budget, manpower, and domain) than with organizational goals. The result is a short-term orientation on problem solving rather than long-term planning. Accordingly, innovation can be stimulated only through civilian intervention, major operational failures, or persistent resource constraints forcing the organization to innovate (Posen 1984). Opposing this view, inter-service contenders argue that military innovation can be facilitated internally, without external civilian interference. This is because military organizations are driven by professional ethos to provide security for the state, which stimulates competition between branches of military services to pursue innovation. Change results not from the civilian intervention but from the work of singular military visionaries or mavericks – senior officers with a new vision of future warfare, reform-minded junior officers, and the establishment of new career paths within the organization promoting reformers (Rosen 1991). In short, inter-service proponents argue that military innovation is stimulated by the competition and debate within a service or between the services.

Using organizational/societal theoretical perspectives in case of Israel, South Korea, and Singapore's RMA trajectories, one would argue that while each society has been relatively cohesive, willing to accept economic burdens of defense in times of war, IDF's and, to a lesser extent, SAF's integrated command and unified force structure (army, navy, and air force under a single branch) coupled with their close integration with Israel and Singapore's defense industries has enabled them to implement military innovations more effectively. Notwithstanding the existence of persisting internal political and organizational debates, divided perceptions and attitudes within and outside the IDF, and, perhaps more importantly, the increasing pressures exerted on the IDF by cumulative changes in Israel's civil–military relations over the past two decades spurred in part by the effects of transformation from socialist to capitalist economy, inflow of new generation of immigrants, and emergence of new Israeli media willing to criticize the Israeli government, the IDF has been able to absorb lessons learned from both its operational successes and failures. Yet, its nearly-constant focus on current security issues has created a short-term orientation, which precluded formulating and implementing a long-term vision of disruptive innovation. Meanwhile, the ROK forces have been entrenched in inter-service rivalries for influence and resources, stimulated by the inherent asymmetric force structure dominated by the ROKA, which has arguably resisted a major change. In this context, South Korea has not been able to fully implement its ambitious defense reforms, although its armed forces have benefited from advanced technologies, increasingly provided by the country's DIB.

Arguably, Singapore's 3G force transformation trajectory has been enabled by a multitude of baseline organizational and institutional factors within and outside the SAF. The development of the SAF has evolved in the context of strong, complex political and administrative structures with high levels of legitimacy and competence. Singapore's civil–military relations could be conceptualized in a "soldier–scholar" framework, in which the SAF has served not only as the key pillar of defense but also as a core incubator for future public servants and industry leaders. At the same time, the SAF has served as a primary social integrator through its National Service, which has effectively created a culture of shared responsibility in ensuring national security. Under conditions of rational, efficient, and effective policy making, deeply embedded in Singapore's institutions, governance, as well as DIB, the SAF has been also part of a symbiotic organizational defense "ecosystem" that has linked the users, developers, and producers of select military technologies. In doing so, however, Singapore has not pursued capabilities for R&D and production of major weapons platforms, but has selectively focused on weapon system integration, upgrades, repair, overhaul, and customization of adapted technologies. In short, Singapore's defense ecosystem, underpinned by comprehensive civil–military defense management, has been closely aligned with the country's cohesive industrial strategy and governance, which have supported the development of the 3G SAF.

Culturalist perspectives

Finally, *culturalist perspectives* may explain the variance in military innovation and use of force through differences in strategic cultures – that is, distinctive, consistent, and persistent views on how states and their military organizations think about warfare. According to Dima Adamsky, "different cultures think differently about military innovation and produce various types of doctrinal outcomes from the same technological discontinuity" (2010, 1). In this view, RMA-oriented military innovation is conditioned by different national "cognitive styles" – that is, strategic preferences, perceptions, ideas and knowledge, techniques and professional attributes, and patterns of habitual behavior acquired over time within members of national strategic community. Seen from this perspective, Israeli, Singaporean, and Korean military elites have developed their particular approaches to the RMA differently and calibrated their military innovation trajectories through their respective historical lessons learned, intellectual climates, patterns of thinking, and communication styles within and outside their military organizations. Specifically, cultural theorists would argue that Israel's trajectory of military innovation can be linked to its informal ("low-context"), problem-focused, action-oriented decision-making styles and organizational ethos rewarding bottom-up initiative, creativity, assertiveness, practicality, simplification, adaptation, flexibility, and tactical improvisation. Israel's strategic culture has traditionally placed a greater emphasis on implementation – emphasizing the performance of its "practitioners" rather than theoretical and abstract conceptualizations of "thinkers," which enabled the IDF to streamline its "learning curve" and adapt to diverse operational requirements directly in the use of force. At the same time, however, Israel's deeply embedded "siege mentality," coupled with IDF's preoccupation with immediate security challenges of the day, created an atmosphere of intellectual conservatism and aversion to long-term strategic thinking, which inherently precluded "disruptive" defense transformation. This can be seen in the traditional reluctance to develop theoretical, historical, strategic studies within the IDF. Moreover, new military technologies were perceived as necessary, qualitative force multipliers, but never sufficient to replace the qualitative elements of the human context.

Israel's RMA trajectory can be, therefore, explained through patterns of implementation, experimentation, and speculation, where technological innovation preceded its conceptual adaptation.

In contrast, while Korean strategic culture has been also traditionally shaped by a "crisis mentality" and perceptions of its geostrategic insecurity – that is, as a vulnerable state located at the intersection of interests of regional great powers – in the Korean strategic narrative, the country has been seemingly unable to cope with externally imposed challenges and historical legacies. In the past 100 years, China's domination of the Korean Peninsula until 1910, Japan's annexation from 1910–1945, Korea's partitioning in 1948 followed by the Korean War from 1950–1953, and superpower conflict and involvement in the Korean Peninsula during and after the Cold War have profoundly affected South Korea's strategic thinking and culture. Emerging from the destruction of the Korean War, South Korea's strategic culture has been externally linked to the collective defense mechanisms of the U.S.–ROK alliance aimed at preventing the break-out of another war on the Korean Peninsula. During the Cold War, the United States has practically dominated South Korea's defense planning, combined training, weapons selection, and overall direction of strategic thought. Military innovation in South Korea thus corresponded largely to military modernization – acquiring and emulating select military capabilities, particularly in terms of "hardware," that would mitigate the overreliance and overdependence on the U.S. military (Cha 2001). Internally, however, South Korea's strategic culture has been relatively conservative in pursuing military innovation. In part stemming from traditional Korean cultural traits and language reinforcing authoritarian Confucianist views through values of filial piety, loyalty, seniority, group orientation, and highly formal "high-context" communication style, and in part due to traditional inter-service rivalries dominated by the ROKA, South Korea's military "cognitive style" has been an exact opposite of Israel's. Implementation would not proceed in the absence of "approved" top-down directives; operational flexibility, adaptation, and improvisation would be neither encouraged nor rewarded within the organizational hierarchy. In this view, South Korea's RMA trajectory can be explained through patterns of speculation and experimentation, but relatively incremental implementation. In other words, one could argue that South Korea's traditional strategic culture precluded greater flexibility and adaptability in translating select defense reforms into practice. Yet, this is also puzzling given the considerable magnitude and impact of the information revolution that has changed South Korea's society and propelled innovation in its economy over the past two decades, but has neither altered traditional organizational force structures nor significantly changed the "cognitive template" of the military. This implies that ROK forces have struggled with organizational change and adaptation – pursuing a cultural change necessary to implement military innovation.

Meanwhile, Singapore's strategic culture has paradoxically served both as an enabler and a constraint in the SAF's 3G force transformation. Historically, the SAF has not been born in battle, nor has it honed a heroic military tradition. In 1967, the political establishment formulated a "Code of Conduct for the Armed

Forces," which has shaped the values and norms for the behavior, position, role, and functions of the 1G and 2G SAF. At that time, Goh Keng Swee, Singapore's first minister of defense, emphasized that in the absence of a professional soldier-caste, a code of guidance was necessary to provide a sense of dignity and purpose to the SAF. The Code stipulated a strictly professional role for the SAF as an instrument of the state in defending its security. It called for all SAF members not to interfere with party politics, remain loyal to the government at any time, be courageous, disciplined, moral, decorous, and dedicated at all times. In 1986, the SAF began to conceptualize a common value system, a process that culminated in a decade later in 1996, with the establishment of seven core values applicable for all ranks within the SAF: Loyalty to Country, Leadership, Discipline, Professionalism, Fighting Spirit, Ethics, and Care for Soldiers. In 2013, the SAF added an eighth core value – Safety. In this context, the SAF has emphasized a high degree of leadership, self-control of behavior though codes of ethics, a meritocratic system of rewards based on intellectual achievement and integrity. Consequently, the SAF has been able to assimilate new technologies and maintain high standards in training, readiness, and combined-arms operations. At the same time, however, Singapore's tightly controlled political and socioeconomic environment has arguably imposed preconceived ideas and solutions in the context of deeply rooted hierarchical structures, processes, and narratives, which have effectively discouraged dissent, divergence, and "spirit of creativity." Singapore's perennial "existential angst" has led to risk aversion if not risk avoidance, both in the civil and military domains. Ultimately, Singapore's strategic culture has precluded an environment supporting individual "mavericks" challenging the established norms through a bottom-up type innovation.

Strategic and policy implications

Notwithstanding their different paths and patterns, Israel, Singapore, and South Korea's RMA trajectories thus reflect evolutionary rather than revolutionary process of change over the past two decades. In this context, the military-technical dimension of their RMA-oriented military innovation has not significantly altered or solved their protracted political and strategic dilemmas. Paradoxically, by improving their defense capabilities through high-value weapons systems, niche technologies, and innovative organizational and operational concepts, the select states have experienced limitations in their use of force by raising the destructive potential and cost of conflict in their regional settings. This is because of the range of exogenous factors, primarily asymmetric responses by existing and potential adversaries. This observation leads to the question of the relative utility, applicability, and receptivity of the RMA-oriented processes of military innovation for other small states and middle powers. While select advanced small states may have the technological R&D base to develop niche weapons systems (i.e., precision munitions, network-centric C4ISR systems, UAVs, etc.) that may increase the range and scope of their military capabilities, the net effectiveness of their respective RMAs will ultimately depend on how well the technologies, organizations, and concepts will be linked to politically defined goals and strategy to achieve them. In other words, RMAs and military innovation must be linked to strategic adaptation, in which national leaders define specific political objectives relative to their security environment, while professional military devise options, concepts, and plans amid prevailing realities of defense resource constraints, organizational, technological, and operational limitations. In the process, small states may translate military innovation into a relative strategic advantage or usable strategic opportunity at least until it will be offset by countervailing responses by opposing forces or new military innovations.

In a policy-oriented perspective, external factors such as changes in global and regional strategic environment are very much juxtaposed by *internal factors* – enablers and constraints, which include available defense resource allocation that shapes organizational and institutional ability to pursue military innovation, and ultimately different aspects of civil-military relations that are conditioned by strategic culture. To begin with, building capacity for military innovation requires relatively high-technology defense knowledge base, access to a range of advanced technologies, precision engineering skills, systems integration expertise, all of which must be propelled by a *sustained investment* in the research, development, evaluation, and testing of military systems and platforms. The fundamental problem for small states with defense resource constraints is the *affordability* of cost-intensive RMA technologies, which must be prioritized amid varying acquisition, training, and operating costs. As a result, financial constraints have serious implications on government options and choices concomitant with the development and acquisition of systems and platforms that shape trajectories of military innovation. The cost factors subsequently stipulate even a greater need for a more efficient allocation of scarce defense resources that must be planned in the context of short- and long-term national security objectives and policies toward ensuring socioeconomic stability and growth. At the same time, however, measuring the "output" of military expenditures alone relative to military innovation paths is bound to limitations. In particular, military expenditures do not adequately reflect the extent of a nation's military capabilities and defense management – that is, the educational and technological proficiency of manpower, the quality and survivability of military infrastructure, the quality of combat research institutions, scope, differentiation, level of sophistication, and degree of external dependence of a nation's DIB. Furthermore, the size and availability of military spending does not sufficiently reflect the *strategic adaptability* of a nation to "convert" its resources into modern military capabilities, in creating and sustaining a force with operational competencies to tackle a wide range of contingencies. In other words, the conversion of available resources into military capabilities is bound to the organizational adaptability, efficiency, and effectiveness of country's *defense planning and management capacity* – aiming to attain better efficiency gains, cost savings, innovation, and competitiveness to maximize the operational performance of its armed forces. RMA-oriented small states such as Israel, Singapore, and South Korea would in theory be able to periodically adjust their defense management processes in order to cope with the ever-changing strategic and operational conditions, resource constraints, while simultaneously developing national innovation strategies to sustain technological edge within acceptable costs and risks. Defense management capacity is essential in reconciling the dual problem of financial constraints in procuring cost-intensive advanced weapons technologies and platforms, while sustaining or upgrading existing military capabilities to achieve national security objectives. It links the many exogenous and endogenous variables that play a role in strategic assessments, military systems development, procurement, and production as well as operations and maintenance, logistical and training requirements in order to reduce the costs, minimize risks, and provide recommendations.

Parallel to defense management capacity, the trajectory of military innovation in small states will depend on the character, level of technological sophistication, and overall development of the DIB – a sector or group of industries capable of specialization in upgrading or modifying major weapons systems as well as developing "niche" weapon component parts and sub-systems, sub-assemblies, and components available for exports. An advanced DIB would be able to adapt to global commercial arms markets by fostering high-technology defense industrial capacity, innovation, and competitiveness through a defense-related R&D–civil research interaction, prioritization, and diversification of defense production. The level of sophistication of a DIB can be conceptualized based on both "hard innovation capabilities" – input factors intended to advance technological proficiency of a country's defense sector; and "soft innovation capabilities" – process-related innovation activities, including organizational, marketing, and entrepreneurial skills (Cheung, Mahnken, and Ross 2011). According to Richard Bitzinger, "[military innovation] demands elemental changes in the ways militaries procure critical military equipment, and reform of the national and defense technological and industrial bases that contribute to development and production of transformational systems" (2008, 7). In the twenty-first century, defense firms are increasingly operating in a complex global environment with increasing competitive economic and market pressures toward "better, faster, and cheaper" solutions while, on the other hand, facing increasing government demands and requirements to supply military organizations with advanced technologies. Understanding the dynamics of the country's DIB thus enables an assessment of the relative sophistication of available military supplies and the robustness of access enjoyed by military forces to a range of innovative defense products. This aspect, however, may present perhaps even more complex problems for small states in terms of *identifying and prioritizing R&D of select technologies, and their subsequent integration paths into military modernization processes, given existing force structures, organizations, capabilities, and current and future threats.* The findings of this study indicate that the pursuit of military innovation by small states challenges not only established roles and mission responsibilities at the operational level but, moreover, broader aspects of defense planning, management, and professional military education at the strategic level, particularly in three key points: identifying and prioritizing between current and future-oriented defense requirements; ascertaining the feasibility, costs, and performance of select advanced weapons technologies; and adopting and adapting new conceptual, organizational, and technological innovations in the operational conduct of armed forces.

Reflecting on these insights, one could conclude that military innovation for small states requires the confluence of three sets of policy imperatives: *facilitating strategic and operational adaptability in defense planning*; *identifying, predicting, and responding to military innovation, whether conceptual, organizational, or technological*; and, in the long term, *pursuing military innovation through changes in strategic culture*. In a long-term perspective, the changing character and complexity of global strategic environment carries a greater level of strategic uncertainty through a convergence of conventional and non-traditional security threats that mitigate the effectiveness of traditional deterrence and defense strategies. Accordingly, the principal challenge for defense planners is developing a "dynamic" range of military capabilities that are agile, flexible, affordable, and multi-mission oriented for a broader range of potential contingencies. This means that traditional linear defense planning processes – that is, assessing environment, identifying specific threats, formulating options, evaluating priorities, and selecting core objectives, strategy, and policy – may no longer be sufficient. In other words, defense planners must facilitate greater strategic and operational adaptability to shifts in geostrategic environment with military forces having the flexibility and robustness to operate in divergent scenarios. In the process, a balanced, functional, and evolutionary approach to military innovation, defense planning, and capability development can be more effective, particularly for small states. At the same time, military innovation is a constant process that must be calibrated in a comparative perspective, relative to conceptual, organizational, and technological metrics of other militaries. The analytical framework used in this study to map military innovation trajectories in the context of evolving strategic thought should help policy makers to make more accurate assessments in this direction. Indeed, comparative perspectives and case studies of military innovation trajectories – whether in terms of "hardware" or "software" – in different geostrategic settings may help policy makers to *detect change* in new approaches to combat and, in doing so, prompt military organizations to debate the validity of established strategic paradigms and operational art. As the ancient Chinese proverb reminds us, "When the wind of change blows, some people build walls, others build windmills."

References

Adamsky, Dima. 2010. *The Culture of Military Innovation: The Impact of Cultural Factors on the Revolution in Military Affairs in Russia, the US, and Israel*. Palo Alto, CA: Stanford University Press.

Bitzinger, Richard. 2008. "The Revolution in Military Affairs and the Global Defense Industry: Reactions and Interactions." *Security Challenges* 4 (4): 1–11.

Cha, Victor. 2001. "Strategic Culture and the Military Modernization of South Korea." *Armed Forces & Society* 28 (1): 99–127.

Cheung, Tai Ming, Thomas Mahnken, and Andrew Ross. 2011. "Frameworks for Analyzing Chinese Defense and Military Innovation." In *New Perspectives on Assessing the Chinese Defense Economy: 2011 Industry Overview and Policy Briefs*, edited by Tai Ming Cheung, 77–80. San Diego, CA: University of California Institute on Global Conflict and Cooperation.

Cohen, Eliot, Andrew Bacevich, and Michael Eisenstadt. 1998. *Knives, Tanks, and Missiles: Israel's Security Revolution.* Washington, DC: Washington Institute for Near East Policy.

Davis, Paul. 2000. "Defense Planning in an Era of Uncertainty: East Asian Issues." In *Emerging Threats, Force Structures, and the Role of Air Power in Korea,* edited by Chung-in Moon and Natalie Crawford, 25–47. Santa Monica, CA: RAND.

Isaacson, Jeffrey, Christopher Layne, and John Arquilla. 1999. *Predicting Military Innovation.* Santa Monica, CA: RAND.

Posen, Barry. 1984. *The Sources of Military Doctrine: France, Britain, and Germany Between the World Wars.* Ithaca, NY: Cornell University Press.

Rosen, Stephen Peter. 1991. *Winning the Next War: Innovation and the Modern Military.* Ithaca, NY: Cornell University Press.

Rosen, Stephen Peter. 1996. *Societies, Military Organizations, and the Revolution in Military Affairs: A Framework for Intelligence Collection and Analysis.* Unpublished Manuscript, June, 1.

Teo, Chee Hean. 2010. "Speech by Deputy Prime Minister and Minister for Defence Teo Chee Hean at the Committee of Supply Debate." *Ministry of Defense,* March 5.

Waltz, Kenneth. 1979. *Theory of International Politics.* Boston, MA: McGraw-Hill Press.

Index